Hacking Artificial
Intelligence

HACKING ARTIFICIAL INTELLIGENCE

*A Leader's Guide from Deepfakes
to Breaking Deep Learning*

DAVEY GIBIAN

ROWMAN & LITTLEFIELD
Lanham • Boulder • New York • London

Published by Rowman & Littlefield
An imprint of The Rowman & Littlefield Publishing Group, Inc.
4501 Forbes Boulevard, Suite 200, Lanham, Maryland 20706
www.rowman.com

86-90 Paul Street, London EC2A 4NE

British Library Cataloguing in Publication Information available

Library of Congress Cataloging-in-Publication Data

Names: Gibian, Davey, 1988– author.
Title: Hacking artificial intelligence : a leader's guide from deepfakes to breaking deep learning / Davey Gibian.
Description: Lanham : Rowman & Littlefield, [2022] | Includes bibliographical references and index.
Identifiers: LCCN 2021062446 (print) | LCCN 2021062447 (ebook) | ISBN 9781538155080 (cloth) | ISBN 9781538155097 (epub)
Subjects: LCSH: Artificial intelligence. | Information storage and retrieval systems—Risk management. | Information technology—Management. | Data privacy. | Data protection.
Classification: LCC Q335 .G53 2022 (print) | LCC Q335 (ebook) | DDC 006.3—dc23/eng20220223
LC record available at https://lccn.loc.gov/2021062446
LC ebook record available at https://lccn.loc.gov/2021062447

To all the amazing women in my life

Contents

Contents

Introduction

Hacking Facial Recognition

THIS BOOK IS A WAKE-UP CALL. INTEREST IN ARTIFICIAL INTEL-
ligence (AI) is at a fever pitch. We are barreling ahead into an
automated future with no guardrails or security guidelines in
place. While many look to the existential risks of AI, such as the
rise of Skynet or artificial general intelligence, the reality is that
we are still far away from those futures. But we are in the midst
of an unprecedented use of AI in operational contexts ranging
from healthcare to military combat, and not enough time, effort,
or thought is being put into ensuring the safety, security, and
integrity of these systems. At their core, AI systems today are like
any other technology. They come with benefits and they come
with risks. Like the automobile, airplane, and software before it,
AI holds significant promise for the human race while simultaneously carrying new risks that need to be understood, mitigated,
and monitored.

This book attempts to provide a framework for understanding
AI risks. This is based on my experiences as a data scientist, U.S.
government official, and AI start-up founder. I have seen firsthand
the promise AI can provide in lifesaving situations but have also
seen the limits of AI quality assurance and security today. From
these experiences, I have come to view AI risk as the defining risk
of our time. And, for the most part, it is not being addressed.

AI risk as a business field began in a bland office in a bland office park in California's Bay Area. The only good thing about the office was the sliver of blue. Perched in the hills west of San Francisco International airport, the office space lacked both the hipster vibe of San Francisco and the glossy feel of the corporate campuses farther south where the likes of Google, Apple, and Palantir are based. The office had been chosen because it fit the two criteria the company had been looking for: It had decent security, and it was cheap. The company's desks were huddled together in a single room at the back of the first floor, away from all of the other offices with an extra set of security locks. The room was cramped and stale, recycled air pumped through the dusty vents. In order to open the extra security locks to enter and exit the room, a whole series of maneuvers had to take place as colleagues would need to stand up, push against the wall, and slide by other colleagues. Sometimes it was easier to just crawl under the desks. But from these windows, on a clear day, a single strip of brilliant blue water could be seen. The San Francisco Bay was visible only through a single gap in the several miles of trees and buildings between the hill and its coast. But regardless, the company liked to brag to potential customers that at only a few months old, they already had an office with a waterfront view.

"OK, so I think we have it," called out Brendan, a lanky Irishman who was one of the most sought-after machine learning engineers in Silicon Valley.

"Ooh, ooh, ooh! Can we put it on the monitor?" asked Neil, the twenty-year-old CEO with a flair for the dramatic. Neil had a vision and had been able to convince the venture capital community that he both knew what the next thing was and could build it. Neil saw a future of ubiquitous deployment for artificial intelligence. And in that future, he saw a currently missing piece of security. Just as traditional software requires cybersecurity to protect it, Neil believed there was enough evidence to suggest artificial intelligence would need its own unique security to defend it against a new set of attacks. The future that Neil saw was autonomous and secure. He had hired the team to build this future with him. And he was growing impatient for results.

"Sure. But in a second. It worked on Friday and at home this morning, but that was only with direct light and when the image was on the laptop display. It's working now with the display again, I made a few changes to the coloring and it's getting better in low light. Let me print out a copy to see if we can make it work a few times without being on a digital display first. I don't like how much the light of the display seems to influence it. Could be something with the pixel configuration and not the image, and we want to isolate the image. If that works, then we can get everyone in the big room," Brendan answered, using the cautious, qualifying language that marked him as a former academic researcher. For him, even big breakthroughs came with qualifications.

Thirty minutes later, the small team was packed into a conference room. Brendan sat at the far end of the conference table, his laptop open with its small camera pointed at him directly. He was wearing a dark blue T-shirt that had no other markings. His image was blown up on the big screen. The only difference was that on the screen, a thin bright green rectangle surrounded his face. When he moved the chair from side to side, the bright green rectangle moved along with him.

The laptop was running an open-source version of a commercial off-the-shelf facial recognition program. At the time, it was not actively looking up who Brendan was, although it could be configured to be run against databases of people and spit out likely predictions. In small, sans serif text at the bottom of the green rectangle was "98 percent," indicating the program's confidence that Brendan was a person. He started the demo by dimming the lights. He then moved his chair around the field of view of the camera. He stood up. He sat down. He turned the lights back on. All the while, the green rectangle followed his face with only the slightest lag time. Brendan centered himself in the camera's field and sat up straight.

He then held up a piece of paper that had what appeared to be cover art for a band on it. It was printed on normal 8.5 × 11-inch printer paper in color from one of the shared office printers down the hall. It looked like it should have had text for a college-town concert on it. As far as design was concerned, it was entirely unremarkable. Brendan moved the paper in front of his chest. The green rectangle around his face disappeared. The paper was not blocking his face. It was squarely in front of his chest, where the logo for a T-shirt might be. But with this unremarkable design, something truly remarkable had happened. He was now invisible to the computer.

"Well, shit," said Tyler, a West Point–trained former U.S. Army officer and the leader of the team's growing national security practice. "This could cause some serious problems for people."

What Brendan and the rest of the team had engineered was nothing short of science fiction. Using a cutting-edge field known as adversarial machine learning, they had created a way to attack sophisticated AI machines with nothing more than a piece of paper. Adversarial machine learning can be thought of as manipulating an AI system to perform the way a third party, an adversary, wants instead of how it is supposed to function. For example, a cyberattacker can follow AI-enabled network endpoint protection systems into thinking that a piece of malware is actually "goodware" because of carefully inserted binary code and letting it onto a network.[1] In other well-known examples, adversaries can force a self-driving car into mistaking a stop sign for a yield sign with carefully placed stickers on the sign that would be imperceptible to a human.[2] Adversarial machine learning can be thought of simply as hacking AI.

Unlike AI research, which has been around since the 1950s, research in the field of adversarial machine learning is only a decade and a half old, having been originally researched in 2004 when academic researchers wanted to understand how to fool spam filters. Research in the field remained shallow until around 2013–2014, when researchers began to understand how computer vision systems could be fooled by injecting noise into the images.[3] By 2019, when Brendan fooled the computer into thinking he had disappeared, adversarial machine learning was a robust research topic with thousands of academic papers being published each year from around the world.[4] Like many digital threats, what starts as a research topic or niche capability can transform into fully fledged cyberattack surfaces as information reaches the mainstream. Practically, this means that as research and development into the field of adversarial machine learning continues to accelerate, attacks on these systems are also going to accelerate. Organizations using AI, whether developed in-house or by third-party vendors, need to be prepared.

Brendan's hacking the AI system with a piece of paper displays a unique trait of this new cyberattack surface. Unlike traditional cyberattacks, where an adversary needs some level of access to the computer or network systems, here the team was able to fool a commercial AI tool with a printed design. No bits and bytes had to be exchanged. For AI systems that operate autonomously or semiautonomously in the real world, such as self-driving cars, in-home audio assistants, military weapon systems, and facial recognition surveillance, a hack on the AI system can happen during the normal data collection process. AI's ability to operate in the real world is what makes it so valuable to businesses, consumers, and governments, and it is also its continual vulnerability.

In the months following the successful hack on facial recognition, Brendan and the team continued to break machine learning–enabled systems. Using the cutting-edge advances in adversarial machine learning, the crack team Neil had assembled broke machine learning systems ranging from facial recognition to audio antivirus software. Some of these attacks were simply the production of research papers applied to real-world scenarios.

Others were built entirely by the team as they pushed past published research onto the cutting edge of this new cyberattack surface. Their successful attacks shed light on an important, and often overlooked, element in the rush toward AI by businesses, the venture capital community, and governments worldwide. AI is hackable.

A few months later, the team was back in the conference room. Some additional venture capital dollars and new clients had allowed it to grow. The scattered team was now dialing in on a videoconference from around the world. Another researcher, Victor, was sitting at the front of the table. He stood up, walked to the back of the room, and turned out the lights. On the screen behind him was a basic display of a few drop-down menus and a large button labeled "Start."

"OK, so we're going to start with something well known this time. We're going with Cryptolocker," he said, referencing the well-known computer viruses that took down hundreds of thousands of PCs in 2013–2014.[5]

Victor had made his career to date as a PhD student focused on robotic control theory, with stints at organizations including NASA and DARPA-funded research organizations. With curly, unkempt hair and a scruffy beard, he looked more like a disheveled Russian novelist than an expert in hacking AI. As the technical cofounder of the company, he was now deeply involved in fundamental research into how and why AI systems can be tricked, as well as how to optimize attacks against these systems.

"So first, so you know I am not full of shit, I am going to throw the file against the system now," he said leaning back in his chair. He then selected "Cryptolocker" from one of the drop-down menus and hit "Run." In less than a second, an alert appeared that said, "Malware—100%."

"See, so it was blocked with a 100 percent confidence score. This basically means that this is a known piece of malware that it recognized directly, which is why it is 100 percent. If we had adjusted it even a little bit, it might say something like 98 percent or 89 percent, whatever. The threshold for this confidence score is well below 20 percent, so it doesn't really matter at these sort of highs."

Victor made a few adjustments to a few additional drop-downs on the display. Critically, he selected two labeled "Random Noise" and "Reinforcement Learning." From another drop-down he selected a well-known, AI-enabled malware classification tool. His actions made sense to the other members of the team but would have made little sense to the general public. He was preparing an AI hack. What was happening on the back end of the simple user interface the team had built was actually quite complex. On the back end of the system was a set of attack libraries and so-called perturbation engines that would manipulate the Cryptolocker file while preserving its ability for operational payload, or ability to inflict harm on another computer. In addition Victor was choosing which algorithms to use to start optimizing this attack.

"And, . . . go!" he exclaimed, hitting the Start button with a flourish. He leaned back and watched.

A new screen popped up with a simple axis. The y-axis was labeled "1–100" and the x-axis was labeled "Number of Attempts." A single dot appeared above the first attempt, right at the 100 line. Over the next few minutes, a scatterplot emerged. As each attempt progressed, the new dots appeared lower and lower down the y-axis.

"What's happening is our machine is learning which noise injections and file manipulations have the biggest impact," Victor voiced over. "Anything that doesn't work is thrown out. We keep that data on the back end, but it gets annoying to show. These are only the successful attempts. The y-axis is the confidence score, which started at 100 percent, remember."

In under two minutes, each new one was coming in at less than 20 percent. When it reached 15 percent, a pop-up alerted the team. "Attack Complete," it read.

"So that file still works, but I don't want to test it on our machines because it's quite literally still Cryptolocker. It would seriously mess my shit up. But I can show it in a sandbox environment later if anyone wants," Victor explained. He was indicating that although they had changed the file, it would still deliver its payload.

What the team had done in this demonstration was successfully hack a different type of AI tool; this time instead of attacking from the physical world they attacked from the digital one. Through a targeted use of adversarial machine learning, they had taken a well-known piece of malware and changed it just enough to be unrecognizable as bad to the AI but would still deliver its payload. Hacking AI in this case could also mean hacking an entire network, if the AI was being used to safeguard it. When AI is protecting a business, personal information, or military secrets, who is protecting the AI?

As I will explain in depth later in this book, the fact that it is hackable alone does not cover all of the risks of AI. Beyond just being hackable, AI is also fragile. It can break. It can be poorly developed. It can be misused. It can be stolen. It can be illegal. It can learn incorrectly. Together, these possibilities for failure are what we call AI risks. These are new risks that are unique from traditional cybersecurity challenges in a number of ways.

AI RISKS

AI risks refer broadly to the performance, compliance, and security challenges introduced by the introduction of artificial intelligence, machine learning, and other autonomous systems into practical use. I'm sure you've seen it. Nearly every day a headline points to the fact that AI is moving out of PowerPoint presentations and into tangible applications. Whether it's through in-home devices like Amazon Alexa, in cars such as Tesla's autopilot, or in military weapons systems, AI has become core to both business and national security strategy.

This book is neither about the benefits of AI nor how to build successful AI strategies. Instead, this book focuses on an often-overlooked hurdle in the successful implementation of AI: risk. This is a book about new challenges that leaders need to be aware of when implementing AI. Like all technologies, AI can fail, it can be poorly developed, and it can be hacked.

We call these challenges AI risk. Unique from cybersecurity, traditional model risk management, and operations risk, AI risks are unique to the nature of AI itself. When talking about AI risk, we don't mean risks such as artificial general intelligence (AI that is autonomously smarter than humans), or the Skynet-style risks of AI dominance, or even the well-placed fears of great power competition in AI. Instead, this book focuses on the practical risks that must be understood, tested, and mitigated in order to take advantage of the benefits of AI. It is our belief that the benefits of AI outweigh the inherent risks it introduces into digital ecosystems. But in order for business and government leaders to operationalize AI in a meaningful way for their organization, they first must understand AI risks and how to mitigate them.

The inspiration for this book came from my experience as a practitioner and start-up cofounder in the AI community. I built AI tools and programs for the U.S. military and the government and later built the world's first AI risk start-up. During these experiences, time and time again I would speak with well-intentioned, highly intelligent machine learning engineers, CEOs, and generals who all wanted to accelerate AI into ever more impactful applications. But when I would bring up the topic of risks, I was met by either a blank stare or laughter. "Of what, robot overlords?" they would ask.

The reality is that AI is not unique from any other technology. Like bridge building, rocket science, and software development, AI can be poorly built and it can be susceptible to a wide range of failure types, including in environmental factors in the real world, adversarial inputs, and bad construction. Today, the world of AI is the Wild West. Few standards are in place, and those organizations who can most benefit from broader applications of AI, such as the financial services, healthcare, and defense industries, are only now updating their model risk management practices to mitigate these risks. This book was written to help guide leaders through how to think past the hype of AI and to understand that while the potential benefits are immense, there are real risks that need to be solved. I believe strongly that in five to ten years, *AI risk* will be as ubiquitous a term as *cybersecurity* is today.

The purpose of this book is to introduce AI risks in a non-technical manner. While some topics, such as model inversion attacks, may be difficult to understand without some background in either data science or machine learning, we attempt to put all topics discussed into layman's terms. When possible, we have provided stories and narratives that draw on our own experience as some of the early hackers of AI technologies. When appropriate, the stories are true. In other cases, we have created future scenarios that mirror today's technical capabilities.

It should worry leaders that nearly every AI system is hackable. The team in the stuffy office with a small view of the San Francisco Bay proved it time and time again, supported by the exploding field of academic research on this topic since 2004.

We know that the team was successful in their efforts to hack AI, because I was there.

A Brief Overview
of Artificial Intelligence

ARTIFICIAL INTELLIGENCE: A HISTORY

To UNDERSTAND WHY AI RISKS ARE SUDDENLY POPPING UP, IT IS important to understand the history of AI as a concept. For all of its recent hype, AI as a theoretical concept is not new. What is new, however, is the explosion in computing power and data availability that has fueled renewed interest in AI over the past decade. Most computer science historians and hobbyists trace the beginning of AI as we know it today to the Dartmouth Summer Research Project on Artificial Intelligence in 1956.[1] Over the last six decades since that summer conference, there have been several different schools of research and academic thought around how to successfully build AI systems. The first school of thought, now known as classical AI, was the first attempt. Later, modern AI developed as additional computing power and the availability of data allowed for previously impossible computations.

The difference between the two schools of thought can be boiled down to control over the system. In classical AI theory, the autonomous system relies on a set of runs. When building AI, researchers attempted to represent the world in a set of defined data structures, such as lists, sets, or trees, and then defined interactions within data structures as a set of rules, such as if-then, and-or, if-then-else, and others.

For classical AI, imagine a photo of a sloth. The programmer would initially seek to describe the essence of sloth, such as "hairy with long arms and legs" and then try to program recognition functions such as "find legs," "identify hair length," and others. The programmer would then explicitly tell the computer program to do more complex problems, such as "find the edges" of the animal.

For early researchers, classical AI theory enjoyed a certain amount of success. Researchers believed that by breaking down the world into smaller and smaller problems and tasks, eventually a robot could interact with the real world with abilities similar to those of a human. Illustrative of this enthusiasm is the famous prediction in 1967 by Marvin Minsky that stated that "within a generation . . . the problem of creating 'artificial intelligence' will substantially be solved."[2] But classical AI, and the hype around it, soon stalled. The world around us is simply too complex for every interaction to be broken down into a series of predefined rules. While classical AI performed with stunning results in carefully controlled experiments, the ability to translate these experiments into real-world business, consumer, or government applications was limited. This ultimately would lead to multiyear-long trends of disappointing results leading to lost research funding. These periods are known as "AI winters."[3] During AI winters, progress in AI theory did not come to a complete halt, but funding was scarce, results were limited, and much of the progress was related solely to theoretical, as opposed to practical, results.

Of course, interest in AI as a theory continued to captivate both academic researchers and, perhaps more importantly, science fiction and film writers. From droids in *Star Wars* to Isaac Asimov's Robot series and everything in between, interest in the possibility of AI is part of what led to continued interest for generations of computer scientists and software developers. But in order to turn these fictions into reality, new approaches were needed.

Modern AI theory takes the opposite approach to classical AI. Instead of trying to define the set of rules, data schema, and logic a computer would follow to make decisions, modern AI relies instead on letting the computer make its own inferences based on data, thereby creating its own rules and logic. This is accomplished by feeding the computer either labeled or unlabeled data, and allowing the machine to come up with its own interpretations.

Under modern AI practice, instead of describing in increasingly minute increments all the elements that make up a sloth in an image, for example, an AI engineer instead feeds thousands upon thousands of images of sloths into the computer with the intent that the computer will extract those relevant features and learn for itself what a sloth is. Another name for this approach is machine learning.

Since 2012, most of the research and practical applications of modern AI have been focused primarily on machine learning, which is a subset of overall AI theory. This branch has proven to be extremely useful in making classifications and predictions based on large amounts of data inputs but requires massive computational power.

As a field distinct from classical AI, machine learning took off in the 1990s, primarily due to its ability to solve practical, as opposed to theoretical, challenges. This in turn drew much of the research focus away from symbolic and logic approaches to AI and instead borrowed heavily from statistics and probability. And then in 2012 a turning point happened that dramatically moved research in AI firmly into the modern AI camp. That year, an artificial neural network called AlexNet beat other image classifiers by over 10 percent. Previously, artificial neural nets were thought to be little more than research tools but were not considered especially practical. But AlexNet's success showed that this approach could beat other image classification techniques at a wide enough margin to be interesting, and research dollars started pouring in. Neural networks remain one of the primary elements of machine learning today.

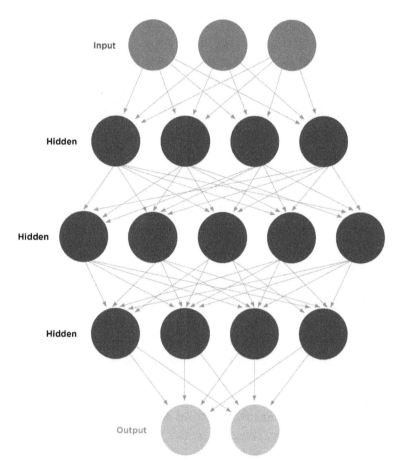

Figure 1.1. **A graphical representation of an artificial neural network with three layers: an input layer, a hidden layer, and an output layer.**

What is most fascinating about AlexNet's victory in 2012 is primarily the age of the technique. The theory behind artificial neural networks had existed even before the 1956 Dartmouth event considered to be the birthplace of AI. This approach was first theorized by Warren McCulloch and Walter Pitts in 1943 in their work "A Logical Calculus of the Ideas Immanent in Nervous Activity."[4] But between the time of their theorization and 2012, important shifts in the computational landscape had occurred that allowed their theory to move into practical application.

First was the data explosion. Modern AI requires a significant amount of data in order to be effective. The more data each machine is fed during the training process, the better the results will be. Before the 2000s, large datasets that would satisfy the underlying requirements of data-heavy modern AI were hard to come by. However, the rise of ubiquitous computing and the Internet led subsequently to the rise of massive datasets. Meanwhile, distributed labeling platforms, such as Amazon's Mechanical Turk, allowed more and more data to become labeled, thereby further accelerating the volume of modern AI input data.

In tandem, Moore's law and distributed computing, including cloud computing platforms, led to significantly more computing power. This computing power helped fuel modern AI's need for massive data to be crunched in order to make results. Meanwhile, cheaper and more abundant supplies of graphics processing units (GPUs), which are more efficient than central processing units (CPUs) for modern AI in practice, also contributed to greater efficiency in the massive sets of computations that need to take place for modern AI to work effectively. The rise of datasets, cheap computing, and cheaper GPUs led to what is now known as deep learning, which is where much of AI theoretical research and practice take place today. The reason why deep learning has been of interest since 2012 is its unique feature that these networks do not saturate when given massive amounts of data. Instead, they continue to learn and improve.

Most of the topics in this book are focused on machine learning, as opposed to classical AI or other areas of AI theory. However, other subjects in the book, such as the need for new approaches to model risk management, should be applied regardless of whether machine learning or any other indeterministic autonomous system is in use. The lessons of this book can be broadly applied to AI risks broadly, although we focus on these

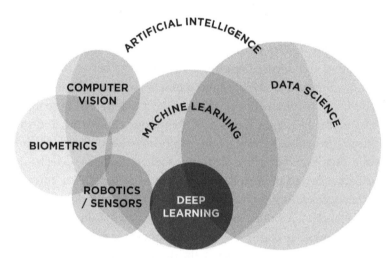

Figure 1.2. Deep learning (DL) is a subset of machine learning (ML), which in turn is a subset of artificial intelligence (AI), the broadest category of techniques aimed at modeling intelligence.

areas to the current interest in and volume of this particular AI application.

The rapid switch from classical to modern AI, and the nearly ubiquitous application of machine learning to data science challenges, is partly to blame for the parallel rise in unaddressed AI risks. Now that AI is not based on rules programmed by the developer, but instead inferred from computational interaction with massive datasets, new risks that did not exist in classical AI theory, such as evasion attacks and model drift, are suddenly emerging. In the eight years since AlexNet proved that artificial neural networks were practical, there has been an explosion of research and development into machine learning and its descendants. During this time, most of the efforts have been forward looking, with researchers and practitioners racing ahead with these tools. Little attention has been paid to the risks of these approaches, which has left potential users of AI with a limited understanding of the downsides of the technology.

In part, the lack of attention paid toward the risks of AI makes sense. There simply was not enough of AI in usable applications for these risks to be anything more than an interesting research area. But today, AI is being incorporated into autonomous vehicles, weapons systems, and consumer devices at an ever-increasing pace. This has flipped the challenge from a research quandary into a realistic operational risk and potential threat vector for bad actors.

The rapid rise of machine learning as the primary AI technology in contemporary data science is partially to blame for unmitigated AI risk. However, the good news is that these risks are known and can be mitigated with appropriate effort.

CAN WE START WITH SOME DEFINITIONS?

Before we begin, I'll tell you a quick story about the definition of *AI*. From 2017 to 2019 I sat on the White House Subcommittee for Artificial Intelligence. This subcommittee, like most, was more impressive in name than outcomes, but it still was able to bring together some of the top technical minds in the U.S. government who were working on, or at least had an interest in, AI. We met monthly, packing into a small, poorly ventilated room on the top floor of the White House Conference Center, just across the street from the famous residence itself. Professionals from all over the government, including the intelligence community; the departments of defense, energy, and transportation; and the lesser known National Institute of Standards and Technology (NIST), among others, met each month to discuss AI efforts and areas of coordination at the federal level. Over the course of several successive meetings in that stuffy conference room, one topic continued to be on the agenda: the definition of *artificial intelligence*.

Without fail, someone near the beginning of the meeting would ask, "Can we start by defining exactly what we mean by AI before we jump into it?" At least fifteen to thirty minutes were spent each meeting debating, discussing, and refining definitions. By the time the subcommittee was renamed and shuffled into a different organization, a definition was still not unanimously agreed upon.

The members of that subcommittee were well-intentioned public servants. Many of them had technical backgrounds and served in technical roles for organizations that maintain high degrees of AI sophistication. In fact, some of the organizations represented, such as the Defense Advanced Research Projects Agency (DARPA), had helped pioneer research in the field dating back decades. But they, like many other highly technical practitioners, struggled to adequately define the terms *AI, computer cognition, intelligence machines, machine learning, deep learning*, and others.

The reason for this debate was not pure academic discussion. Many people feel strongly that key application areas, including machine learning, deep learning, and neural networks, should not be considered true artificial intelligence.

But, from our point of view regarding AI risks, you can forget about technical definitions when it comes to AI unless you are a researcher. For the readers of this book, we propose an operational definition of AI. That is, AI is any computer system that operates in a way that it is not explicitly programmed to operate. For readers who want to go one step further, we like Poole, Mackworth, and Goebel's 1998 definition: "any device that perceives its environment and takes actions that maximize its chance of successfully achieving its goals."[5]

While much of this book technically covers the field of machine learning and its descendants such as deep learning, using these loose, operational definitions of AI will keep us above the minutia and turf wars inherent in a battle for definition of a rapidly evolving term.

How AI Is Different
from Traditional Software

ONE OF THESE THINGS IS NOT LIKE THE OTHER

TO UNDERSTAND AI RISKS, IT IS CRITICAL TO UNDERSTAND HOW AI is different than traditional software. These differences are the reason why proven cybersecurity and secure software development techniques cannot simply be slapped onto AI.

At its core, AI is more complex than traditional software. This is true at every stage of the development and deployment cycle. For simplicity, I have broken this cycle down into four primary parts: primitives, development, debugging, and deployment.

The added complexity of AI versus software starts all the way at the foundational building blocks of AI and software, known as primitives. A common definition of software primitives is that they are the simplest elements available in a programming language.[1] These primitives in software are, usually, the smallest unit of processing and generally consist of a single operation such as copying a byte or a string of bytes from one location to another. It is easy to think of primitives as the foundational building blocks of software, such as bricks on a house or cells in an organism. For software, primitives are just the code used.

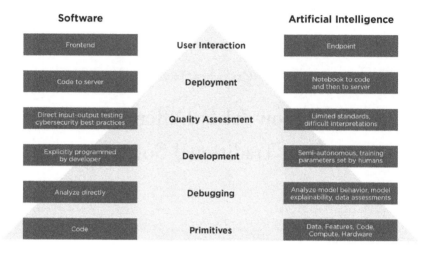

Software		Artificial Intelligence
Frontend	**User Interaction**	Endpoint
Code to server	**Deployment**	Notebook to code and then to server
Direct input-output testing cybersecurity best practices	**Quality Assessment**	Limited standards, difficult interpretations
Explicitly programmed by developer	**Development**	Semi-autonomous, training parameters set by humans
Analyze directly	**Debugging**	Analyze model behavior, model explainability, data assessments
Code	**Primitives**	Data, Features, Code, Compute, Hardware

DEVELOPMENT LIFECYCLE
Figure 2.1.   The differences between software and AI.

AI, on the other hand, has significantly more primitives. These include the data used for training, the features of the data inputs the model uses to make inferences, the code used to build the models, the time needed to train the model, and the computing cost and effort needed for the model to learn and continue operating in the real world.

Following primitives, the development of AI systems is additionally more complex than software. I use the term *complex* here carefully, as many software programs and platforms can be significantly more complicated than AI models. The added complexity is introduced due to the autonomous nature of AI combined with the interaction of the model with the training parameters set by the developer. The result of this interaction is that the developer may not always predict exactly how the AI will respond to given inputs. In fact, if the problem could be solved by logic or decision trees alone, an AI tool would not be necessary. This complexity can make it challenging to understand or interpret AI's actions, while also adding to the challenge of validating the quality of AI models against each other.

When something goes wrong in traditional software, the debugging process can be a painful, confusing process. This is especially true with large, complicated software systems with many dependencies and feedback loops. But the process gets further magnified with AI. This is because while software has a single section to debug (the code), AI has considerably more elements that could be incorrectly or improperly affecting the outcome of the model.

For example, let's compare debugging software versus debugging an AI system. S. Zayd Enam, when at Stanford, created an intuitive set of graphics that we will use to illustrate the processes.[2] Starting with software, when you attempt a solution to a problem or task and it fails, you have two dimensions along which to

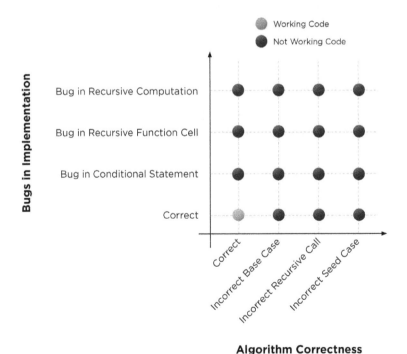

Figure 2.2. Debugging software.

assess your performance. These can be thought of as two axes, with algorithm correctness on the x-axis and errors in implementation on the y-axis. When all of these check out, the algorithm works. When one does not, the code will fail. Thereby, checking to see what is not working with a given algorithm follows a logical, linear process of checking each variable within the so-called search space for the bug.

With AI, however, the situation becomes exponentially more complex. This is because two additional dimensions, the model ideal and the data used, are added. Bugs in the model itself can include the correct features and parameter updates or mistakes in the model selection itself. For example, an AI developer might use a linear classifier when the decision boundaries of the data in question are nonlinear. Likewise, bugs in the data can include data bias messy or incorrect labels, mistakes made in the preparation of the data, or not enough data.

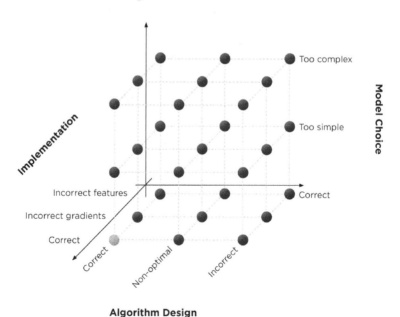

Algorithm Design

Figure 2.3.   Debugging three axes of AI.

It is challenging to visualize all four axes, so instead we can imagine a cube and then a sequence of cubes over time. These visuals should help point to the fact that AI is exponentially more challenging to debug due to introducing n additional variables into the system, which creates $n \times n$ more ways something can go wrong. So with the four dimensions of AI, there are $n \times n \times n \times n$ ways an AI model can fail.

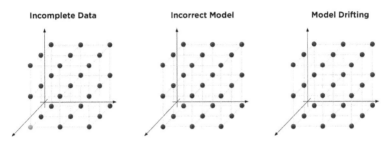

Figure 2.4. Debugging four axes of AI.

An additional factor to consider is the time it takes to train a model. Unlike software debugging where error signals and developer feedback can be quite rapid, AI models can take hours or days to train depending on the data. The same is true of testing the performance of the model. Therefore, when adjustments are made, the feedback loop between error identification and error solution can be quite long.

The differences between software and AI also mean that standard secure software development life cycle processes and development operations (typically called DevOps or DevSecOps) cannot simply be applied on a one-to-one basis. AI developers and organizations implementing AI require the development and standardization of secure AI development life cycles to effectively manage both the added complexity and increased risk of bugs in AI compared to traditional software.

Leaders concerned about AI risks must place a heavy emphasis on developing rapid and effective debugging processes for their

AI. These bugs can range from the inconvenient, such as incomplete sampling of a dataset for training, to the dangerous, such as inadvertently leaving the model open to adversarial attacks.

STRESS TESTING AI

A critical but often overlooked part of an AI development team's workload when assessing AI risks involves stress testing the AI. While this should be part of an AI developer's or data scientist's workflow already, in a surprising amount of cases stress tests on data outside the original training bounds and distribution are not taken into account. Stress testing the model can help explain how the model might fail not by an attacker but instead due to failures induced by naturally occurring environmental factors. For example, self-driving car AI systems must be stress tested in rain and snow simulations, while stock trading AI must be tested outside conditions of normal market volatility, trade volume, and pricing.

It is possible that the AI team will find that failures occur at a high rate in conditions outside the training data. There are two possible outcomes from these failures. First, the team can decide to collect more training data or create synthetic data to retrain the AI. This should be done if there is a likelihood the AI will encounter these conditions in the wild, such as rain or snow. On the other hand, the AI development team may instead choose to develop either a multimodel AI system or a human in the loop system. In each of these systems, when the AI encounters environmental conditions it is known to be poor at handling, the AI is kicked offline and either another model or a human is brought in. Designing AI systems in this manner allows for greater flexibility and operational security in using AI. Although model stress testing is usually done in training, an AI red team should take the view of trying to find situations in which the model fails in order to have the most accurate picture of model performance.

Data Bias

SEXIST MACHINES

"AMAZON HAS SEXIST AI!" THAT HEADLINE ALONE WAS SURE TO grab attention. On October 9, 2018, Reuters ran the headline "Amazon scraps secret AI recruiting tool that showed bias against women" on its online newswire.[1] This was one of dozens of similar articles along the same vein published that day. Over the next several weeks, this story of sexist AI was covered in news articles, technology blogs, and industry publications. Even years later this Amazon failure is one of the most commonly cited AI failures. This leads to the question: How can an AI be sexist?

The simple answer is that the AI learned from its creators. At the core of the problem was a basic premise that an AI engine had internalized the bias of its creators without them knowing it, creating a sexist machine. This clearly was not the intention of the creators of the AI. Instead, the creators of the AI had tried to more rapidly screen candidates for open roles. On the surface, this application of AI makes perfect sense. Amazon has grown to nearly a million employees, ranging from warehouse workers to software engineers. Its operations are global and its human resources department can be inundated with thousands of résumés for each position. Automated systems, powered by machine learning, could rapidly decrease the time it takes to screen résumés while also targeting those individuals likely to be high performers early on.

A member of the team responsible for the AI told Reuters, "Everyone wanted this holy grail. . . . They literally wanted it to be an engine where I'm going to give you 100 résumés, it will spit out the top five, and we'll hire those."

This "holy grail" recruiting seemed to work at the start. It would scan through thousands of résumés and select those candidates most likely to succeed in the notoriously cutthroat culture of Amazon. But it soon became apparent that the engine was spitting out primarily male candidates. This was not explicitly programmed into the machine. In fact, Amazon attests that its engineers specifically removed gender from the features the model should look at. But still, the machine ended up becoming sexist, promoting the résumés of men above those of women. How come?

The models had been trained on biased data, which included the résumés of candidates over roughly a decade. The models were also trained on internal promotion data, which was supposed to help understand which candidates upon arrival to Amazon would be most successful. However, the technology industry is largely dominated by men, leading the machine to infer that men are more successful in technology. At the same time, with much of Amazon's management and middle management being made up of men, the machine further inferred that not only are male candidates better overall, but they will be the ones promoted fastest.

Even though the machine was not explicitly programmed to look for gender, it found it anyway. Applicants with extracurriculars such as "Women's Varsity Soccer" or "Women's Debate Club" were dinged by the algorithm, as were applicants who had graduated from all-women colleges. The machine found a way to add in bias based not on successful predictions, but instead on the existing, nonexplicit bias in the technology industry toward men. The machine internalized the bias of the organization that created it, even though this was explicitly not programmed.

So What Is AI Bias?

AI bias is faulty or unintentional decision-making by an algorithm due to incomplete, unrepresentative training data. Bias is therefore not necessarily an algorithmic risk. It is a risk caused by the underlying data that is used to train an algorithm. Recall that the primitives of AI include the data, as opposed to just code alone for traditional software. Data bias can ultimately not only affect the performance of an AI system, but in some cases can impact its legality as well. Organizations, especially those in highly regulated industries, must take this risk very seriously.

The problem is thought to have been first identified in 1988 when, mirroring the Amazon sexist hiring bot over twenty years later, the UK Commission for Racial Equality found a computer program biased against women and those without European last names.[2] This program mirrored the patterns of human application reviewers with roughly 95 percent accuracy, however, indicating the problem was not in the algorithmic logic but instead with the behaviors of the reviewers that led to sexist acceptance practices.

Bias in human beings is well documented. Some bias may be harmless, such as a bias against hiking due to childhood injuries in the woods. Other biases, such those against a particular race, gender, or sexual orientation, are so harmful to society that laws have been created to limit the ability for individuals to act on these biases. Bias in human behavior therefore can fall between the innocuous and the illegal.

Because AI is generally trained on existing data that was not curated specifically for AI applications (a topic we explore in the next section), existing human biases or data reflecting historical or social inequities can creep into and inform an algorithm's decision-making process. This can occur even when sensitive or illegal features (variables) such as those relating to sexual orientation, race, gender, ethnicity, or age are removed from the training dataset. Amazon's sexist bot is the textbook example of this challenge. Although the creators were well intentioned, the available data led their machine learning system to replicate, rather than remove, existing human inequities.

Bias in AI systems can also come from flawed data sampling. For example, MIT's Joy Buolamwini working with Microsoft's Timnit Gebru uncovered unusual patterns in AI used for facial analysis.[3] Their research uncovered high error rates in predictions for minorities, with even higher error rates for minority women. When they dug into the underlying reason for these errors, it was found that minorities, especially women minorities, were not represented well in the training data, which in turn led to less well-trained models for these individuals and higher error rates.

WHY IS BIAS AN AI RISK?

What can go wrong when the data is biased? There are two primary concerns when dealing with biased data or unrepresentative samples. First, the model can be inaccurate with its predictions, rendering its viability as a practical application limited as well. Second, it can be illegal.

In the first case, AI bias often results in the machine making bad predictions. Take for example a self-driving car. Oftentimes these vehicles are trained in part by using data collected over years from dashboard cameras mounted on human drivers. The vast majority of the data captured is going to be on roads that are generally navigable. When performing on these roads, even in poor weather conditions, self-driving cars have tended to perform well. However, in other conditions the cars fail.

When I worked the U.S. Department of Transportation's Intelligence Transport Systems Joint Program Office (ITS-JPO), we started a project to coordinate the safe acceleration of driverless cars. We took on this project largely for public safety reasons. It is suspected that 94 percent of highway crashes are due to human error.[4] At a closed-door session with major automotive and technology companies (the results were later published) we conducted a voluntary poll of autonomous vehicle companies to uncover what they felt to be the biggest barrier between their technology and safe implementation on a road. The result was surprising. We expected responses such as coordinated federal and state regulatory regimes or advances in AI. Instead, it was a collection bias in the data. We found that the number one concern of these companies when it came to the acceleration of their technology was an unusual one: work zone data.

Self-driving cars were terrible at navigating work zones. This is due to the wide range in work zone signage, configuration, and driving patterns. With each work zone in the training data being unique in some way, the AI powering these vehicles was not well equipped to navigate. This led the U.S. Department of Transportation (USDOT) to create a voluntary data exchange for work zone data, originally called the autonomous vehicle data exchange (or AVDX), now going by DAVI (Data for Automated Vehicle Integration).[5] This voluntary exchange was originally designed to allow autonomous vehicle practitioners to exchange work zone data, thereby leading to more representative training data available and increasing the efficacy of the automated systems responsible for vehicle navigation.

The underlying bias against work zones in this case was not based on human error. Instead it was simply due to the fact that work zones are rare compared to normal driving conditions. The same is true for other automotive edge cases, such as driving in exceptionally bad weather or off-road conditions. But while this data bias did not come from human bias, the undersampling of data in these conditions can have real-world safety impacts. Imagine a self-driving car crashing into a work zone simply because it had not encountered a similar scenario in the training sample. The impact of such an event would likely have lasting implications not only for the manufacturer or algorithm creator, but also for the regulations of that entire industry.

In the second case, bias can be illegal. Take for example the accidental redlining by AI systems. In 1968, the Fair Housing Act outlawed redlining, which refers to the commonly accepted practice among banks to not lend to businesses and consumers within certain sections of a city. Banks would keep their own maps with "redlined" sections drawn on, indicating where they would not lend. These redlines typically followed racial boundaries, thereby preventing minorities from obtaining credit.

Today, AI for lending, insurance, and other financial instruments can take into account many variables, including social media and spending patterns. This data is bound to have correlation with race and other illegal variables such as gender and sexual orientation. For example, Fast Company Admiral Insurance, Britain's largest car insurance company, planned to launch firstcarquote. This AI-enabled system would base its insurance rates on data reviewed by an AI system to include Facebook posts, word choices on social media, and likes and preferences on social media.[6] Ultimately, differences in language, spending, and behavior online provide clues into an insurance candidate's race. AI models have been shown to start becoming racially biased when fed this type of information, creating redlines around different communities even when race is explicitly removed from the data.

Bias of this type can be severely damaging to the people and communities affected. Take for example criminal predictions. Across the United States, many courts use predictive analytics, known as risk assessments, to predict who will commit a future crime. Fantastic investigative journalism by Jeff Larson, Surya Mattu, Lauren Kirchner, and Julia Angwin writing for ProPublica uncovered racial bias in criminal sentencing when using one of the more widely used risk assessment tools in the nation, Northpointe's COMPAS software.[7] The researchers found that beyond not being very useful, only predicting future crimes within a coin toss of probability at its best, the algorithm also predicts that black defendants were 77 percent more likely to be pegged as at higher risk of committing a future violent crime and 45 percent more likely to commit a future crime of any kind, even by persons with similar criminal backgrounds including prior convictions.

The reason for the difference in prediction between minority and white individuals can be traced back to existing biases in the underlying data. For years, police forces across the United States have prosecuted minority communities at heavier rates than white communities. With race also being closely aligned with indictors of criminal activity, such as poverty, joblessness, and social marginalization, the AI learned incorrectly that black individuals were more likely to commit crimes even when all other factors were normalized. The damage of this AI bias, to the individuals, families, and communities affected, is untold.

Both performance bias and illegal bias can create doubt in the predictions of an AI model. If an AI is known to be biased against minorities in sentencing, not hire women, or crash into work zones, it will not be used. Meanwhile, even just the possibility that an AI is biased can sow seeds of doubt into the public as to the efficacy, safety, and ethics of the AI system itself. Until leaders provide transparent reporting into the underlying data to train the model and can show outcomes that do not have biased results, it is unlikely that AI will meaningfully move into consumer and public-facing functions.

What Everyone Gets Wrong About Data

In order to remove underlying data bias to the extent possible from an AI system, leaders and practitioners need to learn to think about data differently. For starters, it is not an input alone. It is the output of a collection process that, most likely, was not designed for AI use from the start.

Let's look at an example of how data science projects often get started at a large organization. A large retailer, for example, might have decades worth of information ranging from customer demographic and spending data, to store-level sales trends, to supply chain information all stored in various databases and formats. The CEO might ask her subordinates to "use those expensive data scientists we hired to really increase sales this year around the holiday season." The data science team would then start to sort through trends, make predictions, and come up with hypotheses around what could drive sales. Given the size of the data, the team will likely use AI to find patterns they couldn't see.

Or in another example, imagine a military intelligence organization. They have reports on a country of interest dating back decades. Information ranges from satellite images, to classified cables, to analyst reports, to open-source information found in newspapers. In recent years, that open-source information has grown to include social media feeds from individuals living in or connected to that country as well. A general might ask his intelligence officer to use AI to better understand the country's military behavior. The intelligence officer, if she is lucky, has at her disposal several uniformed members trained in advanced data analytics. She in turn tells them to get to work on letting the AI understand the country better than the analysts.

In both examples, the data scientists are likely to find something useful for the CEO and general. And often, especially if this is the first time the organization is using AI, these actions will uncover patterns and predictions that were missed or went unnoticed by human analysts simply given the volume of data. But looking back at the previous examples, including Amazon's sexist hiring bot and racial bias in AI-driven criminal risk assessments, both the CEO and the general have to ask themselves whether these are the "right" conclusions. Data bias is such a significant challenge and barrier to AI adoption in part because unless it is carefully proven to not exist, the possibility of is presence casts doubt on all the future predictions.

In order to mitigate these concerns, leaders who want to start using AI must think of the data they feed to data scientists not as inputs alone. Putting the wrong data into a machine learning system can lead to outcomes that can be damaging, rather than helpful, to the organization.

All data is the result of processes, and most of the time these processes are not designed for AI. Data sitting in databases in large enterprises and governments was originally collected for purposes not related to AI. So simply slapping AI on top of it will not help. The better way to think of input data for AI systems is as an output. I break down the data curation process into eight steps. These eight steps are how data *should* be captured with the intention of using it to train AI models. An entire book could be written about this process and the feedback loops it entails, but for leaders concerned about AI risks it is enough to ensure that your team is following these processes at a high level. The steps include:

First, organizations must consider the desired outcome from an AI activity. Is it increased sales? Is it greater insight into enemy intentions? This is the most critical step in the process, because it forces leaders to think critically about what they want to see in the future.

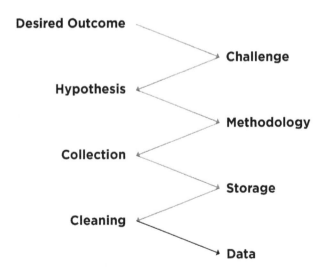

Figure 3.1. Data as the result of a process rather than an input in its own right.

Second, the data science or AI team must define what the challenge between that desired outcome and the current state is today. Is there missing data? Are there blind spots to the organization? This challenge will inform what information should be collected.

Third, the team will create a hypothesis around what data will address the challenge. Data will be defined in the end state, meaning a desired end result of the data itself, complete with ideal metadata information.

Fourth, the team will take the hypothesis and design a methodology to go about collecting this data. Here, data bias needs to be at the forefront of everyone's mind. Does this methodology add unexpected or unwanted bias? Will this methodology result in undersampled populations in the data? Does this methodology magnify existing societal or organizational trends that could have their own biases built in? The methodology design and development is critical to the successful collection of data. Simply relying on prior business or operational practices that were not designed for the AI experiment at hand will likely result in faulty insights and wasted time. The methodology must include the *process* and *standards* required throughout the collection cycle to ensure that only high-fidelity, relevant data is used.

Fifth, the organization has to go about the collection process of actually collecting the data. This may seem straightforward, but collection is where shortcuts are bound to happen. Once while on a project in Liberia researching economic development efforts, a truck carrying the paper results of a survey was involved in an accident. This resulted in roughly 20 percent of our information being lost. The loss of that data was, to me at the time, a great loss. But it was not until we started running the numbers that we realized a greater error. The way in which we collected the data was flawed. There were drastically different answers to certain questions provided by women when the collector noted that men were present during the interviews. This is an example of a flawed collection process whereby exogenous variables can find their way into the data, skewing the results.

Sixth, the organization needs to store the data. This sounds simpler than it is. Many times, AI teams spend significant time and effort simply moving data into an AI-usable format. One U.S. intelligence agency practitioner told me over 70 percent of their data science time is spent moving data around, as opposed to the actual use of this data to create AI models. Storage for AI models must be secure and accessible in order to be useful.

Seventh, the data must be cleaned. At times, this can mean the stripping of illegal or unwanted variables such as race or personal information from the data in order to maintain legal compliance. In other cases, it must be cleaned to remove errors in the collection process. Data cleaning at times can be mundane, such as ensuring the correct spelling or formatting of various inputs. Other times, it can be a delicate balance between data completeness and removing unwanted features or variables from the analysis.

Finally, these steps are complete when the organization has AI-ready data. To implement this process correctly, organizations also need to instill a culture of continual improvement to this process, whereby the organization is consistently evaluating its desired end states and requisite data collection processes to enable the automated insights or systems to get there. Under this operational model, organizations must change their thinking about data from purely an input for AI into a full-blown process with check-ins and stage gates along the way. Such processes are not designed to limit innovation and experimentation, but instead should be considered part of the process for AI success.

How to Limit Data Bias

AI bias tends to fall into one of two camps: incomplete collection or equitable treatment. Sometimes these two overlap as well. Incomplete collection refers to data that is incomplete due to some restriction or limitation on the data collection process such as historical trends, collection costs, or safety restrictions. Equitable treatment, meanwhile, refers to biases in the data that cause the AI to make inferences that take into account illegal or unwanted variables, such as race or gender in hiring predictions. There is a rapidly emerging field of both corporate practice and technical techniques to limit both of these bias types, which primarily fall into what I call synthetic futures and counterfactual fairness.

SYNTHETIC FUTURES

If you don't have the data, why not create it? When data bias is a result of underrepresented samples or poor collection practices, methodologies, or capabilities synthetic data can serve as a stand-in. Synthetic (meaning artificial or generated), data is designed to meet the specific needs or conditions that are lacking from the original data. This artificially created data can help with anonymizing sensitive data in addition to filling in collection gaps.

Over the last few years, the AI industry has turned to synthetic data creation to fill the gaps left by incomplete data collection. Synthetic data engines can be used to create situations, edge cases, and unique scenarios that would otherwise not be captured in collected data. Take for example fraud detection at a major bank. With thousands of transactions and customer interactions a second, humans alone are not able to detect fraudulent patterns. Over the past twenty-plus years, the banking industry has been one of the fastest adopters of automated systems, increasingly using AI to combat fraudsters. However, just as rapidly, fraudsters have learned to adapt. In order to anticipate the future actions of fraudsters, financial firms are creating user profiles and behavior synthetically. This allows them to train the fraud detection system more effectively.[8]

In recent years, synthetic data has moved out of tabular and time series datasets, such as financial trading or fraud detection, and into more complex data including images and sound. Companies selling advanced computer vision capabilities to the U.S.

government, for example, have been creating entire synthetic landscapes to mirror the harsh terrain of Afghanistan. This is done to avoid putting personnel and equipment in harm's way to capture the necessary data to train autonomous systems to operate in the harsh environmental conditions.

Beyond helping to eliminate collection biases in datasets, synthetic data can also save organizations significantly when it comes to the cost of collecting that data. Once the synthetic data engine is created, it can continue to produce fast, cheap data on an ongoing basis. Furthermore, this data generally is created with perfect labels, which saves additional time and resources in the data cleaning process.

Synthetic data is of most interest to regulated or sensitive organizations that must keep certain data secret. This can be due to regulatory reasons, such as personally identifiable information (PII) for healthcare and financial services companies. Likewise, sensitive organizations such as the FBI, intelligence community, or military bodies can use synthetic data to avoid compromising the security of their data or having the data itself be inferred through a privacy attack on the AI. This last point is explored in more depth in chapter 7.

As good as synthetic data sounds, it is not a silver bullet to real data. The process of training an AI model on synthetic data and then applying the AI capability onto real data is known as transfer learning. At the time of this writing, transfer learning remains an exceptionally challenging problem within the AI field. But advances in generating adversarial networks (explored further in chapter 5), are helping to increase the fidelity and transferability of synthetic data as complex datasets are becoming more and more realistic.

COUNTERFACTUAL FAIRNESS

When an AI bias is based on features an organization wants removed, such as race, gender, or sexual orientation, synthetic data will not go far enough. When AI biases are based primarily in the fairness or equitable treatment of persons, careful data preparation must take place prior to training a model. Here, a promising technique is counterfactual fairness. Using this technique, AI developers test that the outcomes of a model are the same in a counterfactual scenario. Practically, this means changing the inputs of race, gender, bias, or other unwanted variables and ensuring that the outcome from the model remains the same.

Silvia Chiappa of DeepMind has been using this technique to help solve complex, sensitive cases such as racial and gender discrimination.[9] This path-specific approach is helpful to untangle the complex, path-dependent relationships human institutions and existing data have between sensitive variables and outcomes. For example, this approach can be used to help ensure that the promotion of an executive into a role was unaffected by the applicant's gender while still allowing the company's overall promotion rates to vary by gender should women-gendered individuals apply to more competitive roles in the firm.

Inherent in counterfactual fairness data preparation is a need to remove those variables in the data that might otherwise act as proxies for the unwanted variable. For example, race might be highly correlated with an individual's zip code or last name. When testing counterfactual fairness, all variables in a dataset must be tested against the unwanted variable and either removed or also tested again for fairness prior to use.

COMBATING BIAS IS AN ETHICAL ISSUE

The ethics of artificial intelligence is closely associated with AI applications for good reason. We don't want underlying human biases, including inequitable treatment, to spill over into our AI. The problem is so pervasive that in February 2020 even the Pentagon issued AI ethical guidelines, which include a significant focus on limiting data biases.

Speaking in 2020, the Pentagon's chief information officer, Dana Deasy, told reporters, "We need to be very thoughtful about where that data is coming from, what was the genesis of that data, how was that data previously being used and you can end up in a state of [artifical intelligence] bias and therefore create an algorithmic outcome that is different than what you're actually intending."[10] Major technology companies have also issued statements condemning bias in AI, vowing to work toward an AI future without bias. But much work remains in this area.

AI bias is a serious risk to organizations that are looking to accelerate AI. Data and its underlying bias is the result of path-dependent processes of creation and collection. And AI today is not equipped to understand these human nuances. While it may be easier to ignore biases in AI, doing so would have disastrous consequences. If AI is understood to be biased, it will erode the public's trust in the system. Leaders must address data bias head-on and provide transparency in their AI training and counterfactual decision-making.

Hacking AI Systems

WANT TO SEE A TANK DISAPPEAR?

HAVE YOU EVER HAD TO HACK A MILITARY SATELLITE TO SEE AN aircraft carrier disappear? Or have you ever wanted to hack an in-home voice system like Amazon Alexa or Google Voice? If these are not enough and you want something a little bit sexier, how about hacking a self-driving car?

Now, what if I told you could reprogram these and other AI systems to do not what they were supposed to do, but what you want. And you can do it not from a computer terminal, but from the real world, without leaving a digital trail. This is not science fiction. These attacks are possible today and they are happening with increasing frequency.

The hacking of AI systems is completed through techniques known as adversarial machine learning. This field of research has grown rapidly over the past decade. Back in 2011, some of the first major breakthroughs in adversarial machine learning were published in academic journals. The next year, in 2012, four academic papers were written about the topic. The number grew rapidly to over one hundred articles in 2014. By 2020, over two thousand papers were being published each year. And these are just the academic journals. At the same time, hundreds of off-the-shelf attack libraries are available online. Big companies, like IBM, provide adversarial robustness tools to test for vulnerabilities, while start-ups and defense contractors are pouring investment into the field. But despite these advances, few leaders in business, cybersecurity, or national security are aware of the magnitude of these capabilities. Would-be adversaries and hackers alike are rapidly gaining capabilities while defenses and active prevention measures are lagging. This is primarily due to an information gap between adversarial capabilities and leadership priorities.

As AI technology matures, it is accelerating into more and more mission-critical systems in business and government. On its own, this is fantastic. Greater adoption of the technology will lead to world-changing breakthroughs. But greater adoption comes with greater risk that the technology will be attacked, corrupted, or manipulated to serve the aims of the attacker. AI attacks can be as basic as injecting a few pixels onto an image or putting a sticker on a stop sign. Or they can be sophisticated, multistep processes involving both traditional cyberattacks as well as adversarial machine learning.

The implications and ramifications of these attacks depend greatly on the use case of the AI. Confusing an AI picking cats out of Internet photos to share with cat lovers is unlikely to yield anything meaningful for the attacker. Hacking a self-driving car, on the other hand, and forcing it through a busy intersection can put lives at risk. Even use cases within the same organization can vary greatly. Hacking a fraud-detection AI can allow an attacker to subvert a bank's fraud detection system. Hacking the same bank's marketing bot will neither cause meaningful harm nor provide significant rewards.

Hacking AI works primarily by weaponizing data. Because AI systems need to take in information, data can be manipulated that can subvert, break, or confuse an AI system to achieve a hacker's aims. This takes advantage of nuances in how AI systems learn and how modern AI systems look at new data to make decisions. Because the data itself is weaponized to take advantage of the architecture that underpins AI systems, these hacks are difficult to prevent entirely. This also means that hacks can be extremely inexpensive and relatively simple to carry out. At UC Berkeley, Professor Dawn Song has been studying adversarial machine learning and its rapidly emerging capabilities. "It's a big problem," she says. "We need to come together to fix it."

The fact that AI is now being hacked is not unique to AI. All digital systems start to be a target for hackers and security researchers once they reach a certain maturity and adoption. From the Internet to the Internet of Things, digital technologies have become targets for hackers wishing to disrupt business or government operations, steal information, pose illicit ransoms, and other aims. Currently, the cybersecurity industry preventing these cases from happening is a $500 billion market globally. That AI is now falling under attack is a continuation of the trend of technological maturity attracting an attacker's attention.

While it is unclear exactly how hacking AI will continue to evolve, we can learn from history. By looking at security patterns from other digital technologies, it is clear that hacking AI is only going to accelerate in the coming years. Take for example the PC market. The first PC was released in 1975. This Altair 8800 did not receive widespread public adoption. It was expensive and its functionality was limited. Therefore, not too many people focused on hacking it. Over the next decade, PCs evolved into systems that are recognizable to children today, including the Apple Macintosh in 1984. From 1984 continuing to today, the PC market has exploded. Now it is not uncommon for there to be multiple computers in a single household.

The rapid rise of the PC made it a prime candidate for attackers. In 1989 Robert Morris started experimenting with Unix Sendmail and built a self-replicating piece of software. This software worm replicated onto the open Internet, which at the time had very few protections. This resulted in an accidental denial of service (DoS) attack. Estimates on the high end are that Morris's worm caused upward of $10 million in damages and caused the entire Internet to slow to barely usable speeds. Later that same year, the Staog Linux virus was created. And then it was off to the races. Cybersecurity was a real risk.

The creation of adversarial machine learning mirrored Morris's experimentation. Hacking AI started off as experiments into understanding how spam was getting through spam filters. The first breakthrough involved fooling a simple computer vision system into thinking an image of a panda was not a panda. But in recent years, these attacks have become more realistic, with successful hacks on in-home voice assistants, self-driving cars, and advanced cybersecurity systems. Now it's off to the races for AI hacking.

THE AI KILL CHAIN

In military circles, a kill chain refers to the structure of an attack. It consists of a few basic steps, including identifying a target, deciding what to do about the target, and executing against this decision. Kill chains systems can be extremely narrow, such as two jet fighters in a dogfight, or broad, such as great power competition. Breaking an adversary's kill chain, meaning their ability to respond to threats, is a critical element to successful battle strategy. Knowing what their kill chain is can be critical to successfully degrading an enemy's ability to wage successful military operations.

It is critical for business and military leaders alike to understand the AI kill chain. By understanding how an attacker will operate, we can defend our systems and prevent attacks from happening. Likewise, we can understand how to get inside an enemy's AI kill chain, using these same capabilities and methodologies in an offensive manner to successfully degrade, cripple, and create mistrust in an adversary's AI capabilities.

A commonly used kill chain example is find, fix, fight, finish. I will use an example of a platoon engaging in small-arms combat to illustrate. First, you start by finding the enemy. This can be done through intelligence, surveillance, and reconnaissance assets or simply by locating an enemy through a scope. To fix the enemy, the platoon commander might pin the enemy down with suppressing fire. Next, she can order the platoon to directly engage, fighting the enemy. Finally, the platoon will finish the enemy by eliminating enemy combatants or effectively disrupting their ability to fight, thus ending the kill chain. The AI kill chain is made up of similar parts. It consists of Find, Access, Generate, Fire, Finish, Feedback.

In the first step, Find, an adversary will identify an AI system in use. This could be through active monitoring of a network, through knowledge of a system, or from the company or organization's own press releases about a new AI tool. The goal of this stage is to identify not only that an AI is in use, but to learn as much information about the system as possible. Conducting reconnaissance and active monitoring of the AI, called AI surveillance, is critical during this time.

Next an adversary will attempt to Access the AI. There are three common access types of AI. WhiteBox access refers to when an adversary has complete insight into the AI, including its underlying training data and underlying logic. GreyBox access is when an adversary only has access to information from the AI's endpoint. The endpoint is the part of the AI that picks up information from its environment, such as a camera, microphone, or cybersecurity node. BlackBox is when an adversary has no access to an endpoint and only knows about the AI's use but can get no information from it. Access will determine the type of attacks that an adversary can perform. The more access an adversary has, the more powerful the attack can be.

During the Generate phase, an adversary will create their attack. Depending on the attack vector, such as whether the attack is starting in the real world or the digital world, generating an attack can take time and computing power. Some attacks are generated to be single pieces of data that, when fed to an AI, will cause the AI to break. Other attacks involve feeding information to the machine to see its response. Whatever the attack type, though, Generating is the most sophisticated part of the operation. This is where knowledge and understanding of AI is most critical.

When the adversary Fires their attack, it is the same as firing a weapon. They launch the attack at the AI and hope that it hits the target. In this case, the hope is that they are able to successfully corrupt the logic of an AI with sufficient strength to achieve their goals. If successful, they then Finish, taking advantage of whatever opportunity they were attempting to accomplish. Finally, a good adversary will create a Feedback loop at the end in order to continually learn from their attacks and increase the speed and accuracy of their kill chain.

HACKING A CAR THROUGH A STOP SIGN

One characteristic that sets hacking AI systems apart from traditional cybersecurity attacks is the ability to originate these attacks in the real world. This is due to many AI applications constantly taking in data from their surroundings, such as voice and sound for in-home voice assistants or full-motion video of the environment for a self-navigating drone. The ability for AI systems to take in new information and respond in real time is part of what makes these systems so valuable. It is also a large part of what makes them vulnerable as well. Adversaries and hackers can use the persistent collection capabilities of the AI as an attack vector.

In 2018, AI threat researchers found that they could successfully hack a self-driving car from the physical world.[1] When most people think about hacking a car, they tend to picture a traditional cyberattack such as hacking through a connected Bluetooth device or through the car's communication system. What was unique about the 2018 attack was that the researchers never needed to communicate with the car at all. They hacked it through stickers.

The research team was looking into how attackers might cause real-world damage to self-driving vehicles by using adversarial machine learning. First, they trained an AI system to recognize street signs. One interesting element they uncovered during the data selection and cleaning process was the fact that regardless of angle, most signs are extremely recognizable. The AI system had very little trouble picking out the signs from its surroundings. However, the uniformity of the signs also made the AI system quite fragile. The team took advantage of this fragility and used it against the AI.

Figure 4.1. The stop sign that hacked a car.

First, the team ran a series of tests to understand how the AI could fail. This process involved injecting noise into the images prior to feeding the images into the AI. After running thousands of experiments, the team had a good understanding of what could cause the AI to misclassify a stop sign as, say, a yield sign or as a seventy-miles-per-hour sign. The team then created a set of stickers. When applied in the appropriate pattern, these stickers would fool an AI into thinking the stop sign didn't exist or was another sign altogether. The unique element of this case was that the stickers look completely harmless to a human observer.

A year later, at CalypsoAI, my team and I replicated this and several other attacks. We built engines for targeted attacks, meaning we transformed the stop sign into a specific other sign, and untargeted attacks, meaning we simply caused the AI to misclassify. Other researchers have replicated this attack using flashing lights, painted dots on the road, and other basic constructions. Creating these attacks have become a popular competition at conferences like the hacking conference DEF CON in Las Vegas and university hackathons. The ubiquity of creating these attacks is part of what makes hacking AI so dangerous. Every day, more attacks are created and new skills are learned by students and would-be adversaries alike. But as of today, very few effective defenses have been developed.

Evasion Attacks

SCARING THE SH*T OUT OF A CISO

CIGARS AT 10:00 A.M. ARE NOT USUALLY A GOOD IDEA. NEITHER are a few breakfast beers in a casino. But, then again, it was BlackHat, the annual cybersecurity convention in Las Vegas. Once a year tucked-in polo shirts try to sell the latest and greatest cybersecurity technology to the thousands of conference goers. There are usually a few good talks, bookended by corporate presentations that cost tens of thousands in marketing budgets to put together. After all, cybersecurity is big business. But that day was more important to me because it was the last day of BlackHat and therefore the day before DEF CON. So basically, it was a day to relax, meet as many people in the one day overlap between the cybersecurity professionals in polo shirts and the T-shirts and flip-flops coming for DEF CON, the world's largest hacking convention. Besides, if there is anywhere to have a few beers and a cigar at 10:00 a.m., it's in Las Vegas.

I had been up since roughly 5:00 a.m., meeting with people who were going to sleep after all-night gambling sessions and catching people coming in from the airport. I was talking to the best and the brightest about adversarial machine learning. It was 2018, and not many in the cybersecurity industry knew anything about it. However, I had been speaking to folks primarily in niche fields—former hackers and cyberoperations specialists from the military and intelligence community who had cashed in for private sector checks. Typically, they still worked only in a small niche, without seeing the big picture. The AI teams, they told me, were in a different part of the organization. So I was not too worried that they didn't know much. I assumed that the big bosses, the chief information security officers (CISOs) and their staff, of the large firms would know plenty about it. I couldn't have been more wrong.

While I was having a cigar and a morning Guinness with a person who allows himself to be described only as a hacker, I got an email connecting me from the innovation team at a big bank to their CISO's deputy. He was available to meet in fifteen minutes. Normally, I would have rushed to get up a demo of my technology and worry about presentation. But, this was the last day of Black-Hat. He was likely going to show up drunk from the night before.

I could not have been more wrong. Showered, shaved, and with a starched shirt, a CISO of one of the biggest banks in America looked the part of his former life. Previously, he had been a senior member of the U.S. intelligence community. As far as I could tell, he used to hack foreign governments and make sure they didn't hack the United States. He never talked about it. When the deputy sat down, he denied a beer and got right to the point.

"Hacking AI isn't a thing because it has no practical applications," he started. "I don't care about tricking a camera that a dog is a cactus. So, what's this about?"

Instead of giving him my normal pitch I decided to go right into the demo. I pulled out my burner laptop, made sure my VPN was working (it was BlackHat after all), and logged into my terminal. From there, I explained that I was going to take a piece of malware and get it past a well-known, foreign-owned cybersecurity company's AI detection system. To start, I sent the malware as is. I was running the software in what is called a sandbox. This is a training environment cut off from the rest of my computer and the Internet. Obviously, when I provided the malware to the AI, it was denied.

Then I turned on an attack library. This library helped me automatically create in real-time perturbations to the malware. When put against the AI, I was able to learn what is called the confidence score the AI has that the malware is, in fact, bad. My software then took the changes in this score and optimized the perturbations. These perturbations were done at the binary level and included injecting noise as well as functional elements into the malware all while not breaking the so-called payload, which is the functional part of the malware script. I explained this to the CISO's deputy in real time. He looked skeptical at first, but as the confidence score shrank from 100 percent to 75 percent and then bellow 50 percent he began to look interested. Then scared. Then intrigued.

Within two minutes, my software had run thousands of perturbations and had optimized the binary code injection and sent it against the AI. The AI said it looked like "goodware" and sent it on through. A piece of malicious code had just been sent past a well-known classifier that was in use at enterprises globally. And it would never have been detected.

"Holy shit," said the CISO's deputy. "I never even thought of this. So what you're saying is that AI is a new attack surface."

"Pretty much," I shrugged. I wanted another beer, but he wasn't drinking. I also didn't think he was going to buy my product, so I wanted to get back to the hackers. He was giving off all the signs of an uninterested buyer. Turned out, he was just worried and wanted to call his boss.

"This changes a lot," he said. "Call me tomorrow." He left worried. I watched his starched-collar shirt standing out as he navigated the maze of the Las Vegas casino to the door. I did call him the next day. It had changed everything.

What I had created in the Las Vegas casino was an evasion attack against an AI. Although the end result, subverting an endpoint protection system to inject malware, was the same as traditional cybersecurity attacks, the way it was carried out was new to the CISO's deputy. Instead of using brute forcing through encryption or relying on human error I had instead attacked the logic of the model itself. In essence, I had created a Trojan horse malware. To the AI, it looked benign. But it carried a dangerous payload. While the logic of these attacks is as old as Greeks and wooden horses, the ability to successfully launch them against AI systems is a relatively new and rapidly evolving field of research and practice. These so-called evasion attacks are the most common attacks against AI and are increasingly defining the new attack surface on AI applications. These attacks have implications well beyond cybersecurity, though, and can impact all AI systems ranging from self-driving cars to healthcare automation. In the coming years, evasion attacks will be the primary defining characteristic of AI hacking and risk.

WHAT IS AN EVASION ATTACK?

Evasion attacks occur when information is fed to an AI and successfully fools the machine. To complete these attacks, pieces of data known as adversarial examples are first carefully constructed. The construction process of an adversarial example can either be a one-off (e.g., taking advantage of a known or transferable vulnerability in the AI) or more commonly constructed after a period of AI surveillance and careful iteration. Increasingly, these iterations can be automated and optimized using reinforcement learning techniques, which are methods that allow a computer to generate increasingly "quality" adversarial examples.

When most people talk about "hacking AI," they're referring to evasion attacks. This is because they're the oldest, and most common (both in the real world and in current research), of AI attacks. Even before they were categorized as evasion attacks, adversarial examples were successfully used by hackers as early as 2004. Back then, hackers and spammers pioneered adversarial sample creation by finding ways to fool automated spam filters that were increasingly relying on early AI applications. These spam filters used linear classifiers, a relatively simple type of AI, and spammers soon found that they were also relatively easy to trick.[1] Back then, few systems were actually using AI, so attackers had limited reason to try to attack them.

However, in the years since 2004, the ubiquitous data explosion and cheap distributed computing enabled the rapid rise of practical applications of AI. And of course, hackers soon followed to break those systems. In 2013, Christian Szegedy was working on research at Google AI and (re)discovered evasive samples almost by accident. He was working on understanding how neural networks make decisions, specifically trying to understand how to explain their behavior after the fact. He discovered what he referred to as an "intriguing property" that all neural networks he looked at seemed to possess. All of the AI's, it seemed, were extremely easy to fool.[2]

Szegedy found while trying to understand and explain how AIs make decisions that they could be fooled by extremely small changes in the underlying data. These small changes, called perturbations, can be as small as a few pixels for computer vision systems such as facial recognition, a few lines of binary code in an automated cybersecurity tool, or slight changes in pitch added into an audio file fed to an in-home assistant. The fact that AIs were highly fragile, and therefore vulnerable, grew into the field of adversarial machine learning. And it all started with these small evasive samples.

Evasion attacks today are the direct descendants of Szegedy's initial findings. They involve creating perturbations in the data such that the machine is fooled. What is perhaps most concerning to AI developers and end users of the technology is that Szegedy's "intriguing property" is that it applies across all AI types. Whether the use case is predicative marketing tools or self-driving cars, whether the AI is used in high-security environments or out in the open, and whether it was developed by a big company or a small start-up, all AI systems are currently vulnerable to evasive samples.

However, as we will explore further in chapter 14, this does not mean that all AI systems are likely to be hacked. An attacker has to have the technical skills, the means, and an underlying payout or reason to attack the system in the first place. While plausible attack vectors may be low today, this is primarily due to the still limited amount of AI in use by organizations. But as AI accelerates into more and more applications, it is likely that this attack surface will expand. When it does, evasion attacks will be the primary method of attacking AIs. We are already seeing it happen today.

THE SCIENCE OF ADVERSARIAL EXAMPLES

The ubiquitous presence of adversarial examples should raise alarm bells for AI developers, organizations using AI, and consumers alike. But their presence begs a question: why do they exist in the first place?

The first theory came from Szegedy's original paper on the topic, back in 2014. He and his team theorized that the poor or improper regularization and too much nonlinearity between relationships in the underlying neural network were to blame.[3] Essentially, this theory states that there are always going to be low-probability situations where a model can be fooled because of the distribution of data and the fact that neural networks make decisions that are nonlinear in nature and therefore are hard to predict.

A few years later, Ian Goodfellow (who later went on to pioneer generative adversarial networks called "deepfakes," explored in chapter 10) and his team proposed the opposite. They proposed that it was because neural networks were too linear in their approach. Inside a neural network, the team hypothesized, decisions were made that were purely linear in their relationship to each other. Each linear interaction perpetuated the prior interactions.[4] Therefore, small perturbations to the inputs resulted in slight changes early on in the layers of the network, which accumulates into massive differences at the end.

More recently, Thomas Tanay and Lewis Griffin proposed the most common theory in use today. Their theory, known as the tilted boundary theory, proposes that because models are simply abstractions of data and never fully fit the underlying data perfected, there will always be pockets that are misclassified.[5] Find one of those pockets, and you have an adversarial example. This explanation seems to make the most sense. AI systems are nothing more than interpretations and abstractions of the real world. So unless a training set is complete within the range of possible outcomes, there will always exist examples that the model does not predict with 100 percent probability. It also helps that Tanay and Griffin disproved the other two approaches, which lends credibility to their arguments.

While other explanations exist, including an inherent lack of training data for AI[6] and computational challenges of building AI that is robust against being fooled,[7] a final theory worth mentioning is that adversarial examples are not a bug of AI. No, the authors from MIT argue, they are a feature in how neural networks engage with the world.[8] They argue that while to us adversarial examples are challenging because humans can't perceive them, these researchers flip that on its head and argue that just because we humans are limited with our faulty eyesight and three-dimensional thinking doesn't mean the machine is. What we view as adversarial examples are just evidence of a higher-order pattern recognition by the machine. Of course, even if that's the case, it's not going to solve the fact that a self-driving car can be forced off the road by stickers. So I tend to ignore this theory in favor of ones with more practical applications.[9] However, this same research also demonstrated interesting elements of attack transferability and should be explored by technical readers.

Whatever the reason for their existence, adversarial examples and the threats they pose are not going away. They have been found in every application of AI, including computer vision, natural language processing, speech and sound recognition, time series analysis, predictive analysis, and others. The fact that so much remains unknown about the underlying science of adversarial examples makes them particularly dangerous for AI developers and AI users looking to harden their models against attack. As of today, there is no 100 percent guaranteed method that a model can be defended against attack.

TYPES OF EVASION ATTACKS

Most people have the tendency to look at types of evasion attacks based on usage of AI—for example, computer vision attacks or attacks on natural language processing. However, at CalypsoAI we developed a now-public approach that I believe better encapsulates how leaders and users of AI should think through their AI risks. Because all AI uses have been shown to be susceptible to evasion attacks, it doesn't make sense to go one by one through them as the underlying mechanics will be the same.

Instead, it is better to think about types of attacks as a function of how much access an adversary has to your model and underlying data. This access level will influence the techniques available to an attacker and will also determine the level of risk to the organization. Full access to a model, known as WhiteBox attacks, are the most dangerous, as an adversary has complete information about your underlying model. Next, GreyBox attacks assume an adversary has some information, such as a confidence score, from the model. Finally, BlackBox attacks assume no information about the model and are the least threatening, but also likely the most common.

All evasion attacks also fall into one of two categories based on the specific outcome they are trying to generate. Targeted attacks attempt to shift a model's decision in a certain way, such as thinking a stop sign is in fact a yield sign. Untargeted attacks do not seek to have a defined outcome, so long as it is the wrong outcome. An untargeted attack tends to be more common as it is generally easier to pull off. Evasion attacks across the three primary categories of WhiteBox, GreyBox, and BlackBox can all be either targeted or untargeted, while WhiteBox attacks have the highest probability of being targeted due to the additional information an attacker has during this attack type.

WhiteBox

WhiteBox attacks are possible when an adversary has complete access to your model and the underlying training data. These are the most powerful types of attack because the attacker has complete information about your model, how it was trained, and how the model "thinks" (or at least as much as you do). This allows attackers to craft carefully constructed attacks. Because they need complete information, these attacks are relatively rare today. But this might change in the future as the goal of traditional cyberattacks increasingly becomes AI surveillance.

In a WhiteBox attack, an attacker understands the underlying gradients of your model. Gradients, at a broad basis, are a representation of how an AI makes decisions. The attacker can use this knowledge to create mathematically optimized adversarial samples that will fool the AI with high probability. Research has shown that when an adversary has access to the gradients of a model, she will always be able to craft successful attacks.[10] Because they take advantage of the gradients in a model's decision making, WhiteBox attacks are sometimes referred to as "gradient-based" attacks.

AI surveillance in the context of a WhiteBox attack refers to cyberattackers gaining access to a computer network or system in order to understand an AI model's internal workings and underlying data. Due to the high levels of access needed to successfully pull them off, WhiteBox attacks are typically accompanied by traditional cybersecurity or insider threats designed solely for AI surveillance activities. An adversary pulling off a WhiteBox attack will need access to the network, systems, and databases that are being used to develop the model. These should, in theory, already be well protected by cybersecurity hygiene and best practices. However, we know that this is not always the case, with large organizations ranging from Equifax to the U.S. Office of Personnel Management's security clearance records being hacked in recent years.

In the coming years, cybersecurity breaches will increasingly be used not for data extraction but instead for AI development surveillance. With this knowledge, an adversary will then be able to conduct WhiteBox attacks against a model. Because there are no known defenses against WhiteBox attacks on an AI, an attack will always be able to succeed. For this reason, AI development surveillance is something that needs to be top of mind for cybersecurity analysts and operations centers when monitoring their network. Should an adversary be able to create a WhiteBox attack for a model in use in a mission- or business-critical environment, the results can be disastrous.

Glasses That Hack

WhiteBox attacks are the most powerful type of attack, and no AI system today is fully defendable against them. However, not all AI systems have a clear attack vector, described in chapter 11. However, one of the clearest AI applications that adversaries are trying to hack is facial recognition.

Facial recognition technologies and other biometric identification capabilities rely on AI to crunch the massive amounts of data needed to pick a human out of a scene and then identify this person. The ubiquitous surveillance market is rapidly growing as countries, municipalities, and businesses hope to gain security benefits. In China alone, the number of surveillance cameras in use is expected to reach 626 million.[11] By comparison, the United States has roughly 40–50 million surveillance cameras in use.[12]

On the surface, anyone trying to beat facial recognition must be a criminal, terrorist, or other nefarious actor, right? Why else would they be trying to evade surveillance? But reality is more complex. Human rights activists, political dissidents, and even ordinary citizens concerned about government overreach all have reason to distrust that ubiquitous environment. This is especially true in autocratic regimes, such as China and North Korea, but can also extend to privacy-minded small-government activists in Colorado; Black Lives Matter protestors in Portland; and peaceful environmental activists in London who fear that facial recognition will invade their privacy in unwanted ways.

These many intersecting reasons, as well as the fact that facial recognition provides a highly visual example, has made the hacking of facial recognition systems a primary target for researchers and AI adversaries alike. Over the past few years, the adversarial machine learning community has become highly adept at crafting WhiteBox attacks against these systems, even going so far as to create digital WhiteBox attacks that not only fool the AI into misidentifying a person but also can be easily transported into the real world.

In what has become a famous example in the rapidly growing field of hacking AI, a team was able to fool an AI system using digitally rendered glasses that could be created in the real world. A classical attack on any image-classifying AI system starts by changing a subset of the pixels to understand how changing pixels will affect, either positively or negatively, the performance of the AI. These changes are known as perturbations, while the amount the images are changed is known as the perturbation distance. For several years, the primary way to hack a computer vision AI was to change pixels across the image. However, these were hard to replicate in the real world because large sections of the image are out of the control of the attacker—for example, the background or lighting was perturbed as well as the face. This means that people looking to avoid facial recognition for privacy, activism, or nefarious reasons lacked the means to take advantage of these attacks in an operational way.

This changed when a team hacking facial recognition found a way to limit the perturbations to only a limited shape.[13] These adversarial patches, as the perturbed sections of images became known, restrict the perturbations possible to only areas that could be replicated in the real world. For example, the adversarial patch can be limited to areas on a person's face in the shape of a pair of glasses. Using WhiteBox attacks, the attacking team was able to optimize their attacks using only this limited perturbation space. Once the attack has been optimized, these glasses can then be created in the real world and used to fool the classifier.

The same hacking team that created the glasses adversarial patch is continuing their work to create a universal robust adversarial patch. Universality means that even when models are specifically trained against adversarial attacks, the adversarial patch will still work.

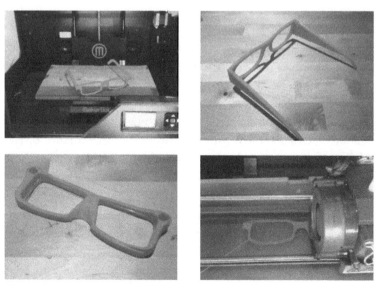

Figure 5.1.   Glasses that hack facial recognition.

Carrying out a WhiteBox attack in the real world is challenging due to the additional cyberhacking skills or other capabilities required to deeply understand all of the primitives of the model required to build the attack. Sophisticated cyberintrusion and AI surveillance is required in order to successfully implement these attacks, requiring a different skill set than the data scientist developers have who hack the AI logic itself. Pulling off a large-scale, long-term cyberpenetration can be technically challenging and expensive, requiring both social engineering and cybersecurity expertise to pull off an effective hack. But it is possible. Given the significant benefits to attackers and would-be attackers of the AI system for personal, activism, or criminal activity, it is highly plausible that such attacks are actively underway for most major surveillance technology companies.

GreyBox Attacks

In a GreyBox attack, an adversary does not have complete access to the underlying model but instead has access to some level of output from the model.[14] These outputs can be the confidence score of a model's prediction or the hard label assigned by the classifier. A confidence score is a probability of 0–100 that an input is a certain input. For example, an image classification tool might think an image is 93 percent a European swallow and assign that label to it. A hard label, on the other hand, is the same label, European swallow, but without the score attached.

A GreyBox attack uses these predictions to continue manipulating their inputs to create better and better adversarial samples. So in essence a GreyBox attack can be thought of as seeing how a model reacts to an input and crafting better and better inputs to try to eventually beat the model. In a GreyBox attack, uninterrupted access to a model's endpoint (meaning the location a model takes in information such as a camera or other sensor) is required. The most important thing a leader should understand about a GreyBox (or BlackBox) attack is that they do not necessarily require any additional cybersecurity hacking skills. These attacks take advantage of AI's need to continually gather new information in order to be useful. For example, a self-driving car's cameras or LIDAR (Light Detection and Ranging equipment) need to be on or a voice assistant needs to listen. These endpoints often have a way to capture some information about their output, often by design.

If it seems strange that an AI would give away information about its predictions, due to the presence of attackers, it is important to remember that this information is shared by design. For example, the developers of a self-driving car want to easily determine what the vehicle is seeing, while allowing insurance companies, regulators, and other interested parties easy access in case of an accident. They therefore may allow you to view the score directly as a developer, or have an easy way to gain access to it even by a third party. In the cybersecurity market, malware detecting endpoint protection platforms (EPPs) often provide a confidence score of 1–100 or 1–10. This allows someone submitting a file or interacting with the system to know that the error is in order to potentially fix it. But these feedback loops developed with benign intent provide pathways for attackers to learn more about how a model responds and craft GreyBox attacks against them.

The most powerful attacks on Greybox models are known by names unfamiliar even to many AI developers, including ZOO,[15] SPSA,[16] and NES[17] for confidence score attacks and the Bourndary Attack method for hard labels.[18] But their relative anonymity today hides not only the fact that these powerful tools are not only widely available for public consumption, but that many attack libraries, generators, and tool kits are being developed for hacking AI using GreyBox methods. I suspect we are just months away from a tool kit equivalent of the Kali Linux penetration testing and hacker tool kit used in traditional cybersecurity testing.

When we scared the sh*t out of the CISO on the casino floor in Las Vegas, we had constructed that attack using GreyBox attack methods. Granted, we had taken it one step further than most methods. When building our attack, we supercharged our GreyBox attack. We employed reinforcement learning methods to rapidly optimize our attack. This allowed us to compress the perturbation trial and error period into just a few seconds. That we were able to carry off this attack against a commercial classifier in only a small amount of time highlights the significant vulnerability organizations using AI have to GreyBox attacks. Most of the examples in this book are GreyBox in nature, and leaders must not only be aware of these risks but must also take active measures to prevent them from happening.

BlackBox Attacks

Unlike both WhiteBox and GreyBox attacks, BlackBox attacks assume zero information is gathered from interacting with the AI system. This makes them the most likely to be attempted but hardest to pull off without significant time and computer capabilities. BlackBox attacks are often known as brute-force attacks. They cannot be optimized, either through basic pattern recognition or reinforcement learning, and therefore they rely simply on changing the input enough to confuse the AI model. Common methods for attacking an AI using BlackBox methods include randomly rotating image inputs,[19] applying lots of common perturbations,[20] and simply adding noise to the input within a Gaussian distribution.[21] These attacks can be quite useful, though, as injecting sufficient noise can greatly confuse an AI and cause it to fail.

While BlackBox attacks may not seem particularly tech-savvy, in many ways they are the most likely attack to happen in the real world. For example, purposefully wearing many bright colors, or painting lines across one's face, or obscuring one's face with a mask can be thought of as BlackBox attacks against an AI. In these examples, the wearer of the disguise is acting as an adversary by trying to trick the AI with limited knowledge of how the model is actually working on the inside. They are guessing, however, that they have altered or obscured their face enough to confuse the machine. The adversary is brute-forcing the data input to the machine to be so different as to alter the classification. Likewise, adding enough noise into a computer file to completely obscure a single malicious piece of code is also a cheap and potentially effective way, when done at scale with enough computer power, to evade AI cybersecurity tools. Other examples of real-world BlackBox attacks on AI systems include pointing lasers to confuse surveillance equipment, adding reflectors on a warship to refract light to confuse a spy satellite, and hiding malicious content in gibberish to evade content censors. In all of these examples, the adversary did not need access to the model's architecture, the underlying training data, or any additional information about the model. All they needed was the endpoint and enough trial and error tests.

Transfer and Surrogate Attacks

Transfer attacks refer to successfully crafting an attack against a known or internally developed model and then using the attack on a similar, externally developed model. These attacks are also called surrogate attacks. You can think of transfer attacks as building a test case or training center for your attack before bringing it into the real world.

The concept of a transfer attack is well known to national security leaders and those who have been involved with special operations in the military. Prior to completing complex operations, special forces teams will create entire mock-ups of compounds or structures that they will encounter during a high-profile raid. This was famously completed by the Naval Special Warfare Development Group (commonly known as SEAL Team 6) during their preparation for the raid on Osama bin Laden's compound outside Abbottabad, Pakistan.

After determining that there was a high probability that bin Laden, controversially thought to have been code-named Geronimo, was at the location, the National Geospatial-Intelligence Agency captured high-resolution photographs of the compound from satellites. As Operation Neptune Spear, as the operation was officially code-named, ramped up, the special forces operators who would be on the operation needed to practice. A complete mock-up of the facility was created at the secretive, government-owned facility named Harvey Point in North Carolina.[22,23] This structure mirrored the architecture of bin Laden's compound in Abbottabad, giving the SEALs as realistic a training target as possible.

Transfer attacks use this same concept. An adversary will try to create a model that resembles their target model as closely as possible. While key internal elements of the model may not be known, an adversary can make their best guess as to the decision-making logic of the AI and then build WhiteBox tools to fool them. In that way, transfer attacks typically start as WhiteBox attacks on internal models but are then applied to additional models. AI security researchers have found that evasion attacks have a high degree of transferability. This means that attacks created based on a WhiteBox attack can be tested against models completing similar tasks with a high degree of success. Why evasion attacks can be easily transferred remains an open-ended research question.[24]

Transfer attacks provide an adversary with a way to take advantage of the power of WhiteBox attacks without the additional cybersecurity capabilities typically required to pull them off. An attacker attempting a transfer attack has a few possible ways to successfully construct an attack, based on their knowledge of and access to the model.

First, an adversary may try to rebuild the model by querying the endpoint of the model. If the AI provides information as part of its output, sometimes referred to as being an oracle, querying enough times will give the attacker sufficient data to reverse engineer the model.[25] Once the model has been reverse engineered, a WhiteBox attack can be constructed and optimized for a successful attack.

If the AI does not provide a query-able endpoint, an attacker can instead try to construct a similar model. For example, if the AI is being used to do a common, or well-known, task such as malware classification, facial recognition, or object recognition, an attacker can create a model based on the same data as their target AI. They would need access to this training data to create this model, but common AI applications often have open-source information available that has become widely used in model training. Unless the target AI is trained on the basis of proprietary or sensitive information, it is possible that the same library a model was trained against is available for others to use. Once an AI model has been used to complete a similar task, the attacker can make an educated guess about the architecture of the model and construct an attack from there.

Finally, if the attacker has no knowledge of or access to the data or model, the attacker can build a model that completes the same task. They can do this by taking an off-the-shelf model from AI development firms or they can build a model using their own dataset. At the outset, this may appear to be a flimsy attack. But evasion attacks constructed in this manner often have a high degree of success when transferred between AIs.[26] The transferability of WhiteBox attacks created on one AI system to other AI systems completing a similar task remains an unsolved, and therefore unmitigated, security concern when it comes to building secure AI systems.

Let's look at how, and why, an adversary could construct a transfer attack. In this example, we will assume that you are a human rights activist network operating in a large city in a foreign country with a history of human rights abuses. You have reason to believe that you and your colleagues will be targeted by local security forces due to ongoing protests that you have organized regarding the country's upcoming elections. This country, like many in the region, has recently added a significant amount of security cameras in the city, creating a ubiquitous surveillance net around government buildings and the business district. These are exactly the places you need to get to. Due to the risks, you want to transit to and from the protest sites and your home, as well as go about your daily life, without disruption. So you decide to fool the ubiquitous facial recognition system without attracting attention.

To target the AI system and create a hack, you need to know more about it. The name of the company that developed the AI is a good place to start. During your research, you find that a large contract for the installation of security cameras and a facial recognition system was not released by the city government or the security forces, but you are able to find out which company provides these by first looking at the brand of surveillance cameras in use and then using Google Search (or, if you're even more privacy minded, DuckDuckGo) for that company plus the name of your city. You find a press release about the contract in their online, publicly available PR archive. Now you have your target AI system.

Next, you will create a similar facial recognition AI. Many facial recognition classifiers are available on the Internet, through open-source libraries, and by companies. Meanwhile, there are many open-source and purchasable libraries of tagged photos and videos available to train models. Through open-source research and by evaluating how the city's surveillance cameras are used, you can start to create an educated guess around the AI logic used in the system, including key features used by the AI to determine identities. Now, it is important to note that your model does not need to be an exact copy of the target model, due to the high transferability of powerful attacks from one AI system to another completing a similar task. After trial and error with many different models, you pick an AI model that you believe closely resembles the target model. Then you get to work.

Because the model has been created by you, there is plenty of information to create a WhiteBox attack. Because you don't want to attract attention, you decide to limit your perturbation area to create an adversarial patch that can easily be created in the real world. You optimize a WhiteBox attack, which gives you as close as possible to a single-pixel attack in the form of a red dot on your cheek. Next, you go down to the local pharmacy, choose a box of children's Band-Aids with colorful designs, and put one in the exact location from the attack. You test the attack back on your model and find it has decreased its confidence score by 75 percent and now misclassifies you as someone else. Success. You have hacked an AI system using a transfer attack.

It must be stated that there are several limitations to operationalizing these attacks. Modern AI systems in use can look at any number of factors in determining identity, ranging from pattern-of-life analytics to the gait of someone's walk. Therefore, these sort of attacks remain challenging to successfully operationalize without serious technical know-how. But there are two trends in favor of the attacker. First, most AI systems in use today are relatively basic and transfer attacks have a high rate of success against them. Second, more and more tools are becoming available to aid AI attackers.

At CalypsoAI, our research team would regularly comb through the Internet, including the dark web, to find attacks against commercial classifiers. In 2019, we found 184 attack libraries against known classifiers available, including against commercial facial recognition. And we found them all as a side project. At that time we had a total team size of between ten and twenty people, and all of us were doing a mix of machine learning research, product development, and sales. I have no doubt that our research efforts on the topic fell well below those of a dedicated, persistent actor or state-sponsored entity. Many of the attacks we found were designed to hack current best-in-breed computer vision models. These attacks were typically named for the AI they were designed to break and therefore tended to have been obscure. For example, one high-quality facial recognition hack refers to its method as LResNet100E-IR with ArcFace loss. For that specific example, we actually did not need to find a hack on the dark web—it was published in an academic journal.[27]

In order to stay relevant in the rapidly advancing field of AI, many companies use best-in-breed models developed elsewhere in their products. The incorporation of these models into products, paired with the rapid acceleration of academic papers and dark web attack libraries to beat these models, means that well-intentioned companies trying to remain at the cutting edge of AI science and capabilities may unwittingly be adding new vulnerabilities into their digital environment—vulnerabilities akin to the sort that our hypothetical activist can take advantage of.

With over two thousand papers published each year by academics and new attack libraries popping up on the Internet every few weeks, the environment has never been better for a would-be AI hacker. And this is only going to get worse as the fiend of adversarial machine learning rapidly moves away from being a trivial academic matter into a full-blown cybersecurity risk.

CHAPTER SIX

Data Poisoning

WHAT IS DATA POISONING?

DATA POISONING ATTACKS ARE WHEN AN ADVERSARY SUBMITS malicious data to an AI system as a way to force the system to behave the way the attacker wants, as opposed to its creator's intent. These attacks take advantage of one of the foundational primitives of AI, the underlying data itself. By submitting faulty information, an attacker is able to shift an AI's behavior or decision-making. Historically it was believed that data poisoning attacks only take place at the training time. This means that an adversary would have to gain access to the dataset being used to train the model in order to inject faulty information. Recent research, however, has shown that malicious activity can be submitted to an AI to force it to incorrectly learn while in use. Attacks that occur while the AI is actively working are sometimes called adversarial drift or online system manipulation.

Figure 6.1 details how data poisoning works. In this example, you can see the classification boundary between two classifications of data, circles on the left and triangles on the right. This two-dimensional classification boundary is a common way to visualize how an AI system working on classification looks. In this example, just a single piece of data is changed. This is represented by the triangle and arrow on the right-hand image moving up

71

and to the left. Due to this shift in the value of underlying labeled training data, the entire classification boundary shifted. In the image, this could have happened simply due to new collection or new data being input.

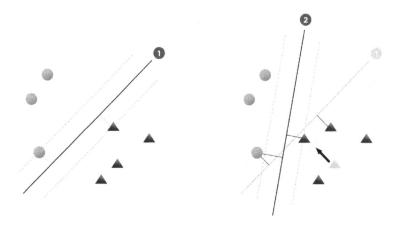

Linear SVM classifier decision boundary for a two-class dataset with support vectors and classification margins indicated (left). Decision boundary is significantly impacted if just one training sample is changed, even when the sample's class label does not change (right).

Figure 6.1. Shifting the boundary through data poisoning.

Data poisoning attacks occur when an adversary feeds specifically selected data into an AI either in training or in use such that these classification boundaries, or other decision-making functions, fail or respond specifically to that adversary's intentions and against the intentions of the developer. These attacks have been successful against facial recognition systems, sentiment analysis tools,[1] malware detection,[2] cyberworm signature detection,[3] cyberattack detection,[4] intrusion detection,[5,6,7] and many others. For readers who are interested in digging into this particular attack type in depth, Ilja Moisejevs's blog post on Towards Data Science on this topic serves as a highly accessible primer.[8]

Similar to the evasion attacks explored in chapter 6, poisoning attacks have been found to be highly transferable between

AI models. This means that once an attack has been developed, an adversary can use it across multiple models, even when the model is retrained to perform a different task.[9] As organizations look to outsource the development of models, or increasingly use pretrained models as part of their workflow, the transferability of attacks embedded in the models originally can spill over into the new tasks. This means that both the developer and the end user of the model might be unaware of the risk.

Data poisoning is typically broken down into two categories: availability attacks and integrity attacks. They both share the same characteristic of using certain amounts of specifically selected or altered data as a weapon to modify the behavior of a model.

Availability Attacks: AI Learning Gone Wrong

Availability attacks aim to provide an AI with such high levels of bad information that the learned behavior is useless. These attacks are also referred to as model skewing, because they are designed to change the behavior of a model in such a way that the model starts to misclassify inputs. An attacker can use these methods to either instill levels of doubt into the performance of a model or to systematically alter an AI's behavior in a way that will benefit the attacker unknowingly to the AI user.

For a while it was believed that only a limited class of AI types, especially binary learning algorithms and support vector machines, were susceptible to poisoning attacks. It was theorized that this was due to the high levels of complexity needed to optimize data inputs for attackers to create an attack. More advanced AI techniques, such as neural networks and deep learning architectures, were thought to be immune due to the more advanced insights these AI types generate and the difficulty of understanding their decisions.

But this is not the case. In 2017, a team of AI researchers was able to hack advanced AI systems using similar techniques applied to the more complex learning of advanced AI. Their research has shown that advanced AI capabilities are, in fact, highly susceptible to availability attacks as well. Neural networks and deep learning approaches used in a wide range of AI applications—including spam filters, malware detection, and handwritten digital recognition—all can be poisoned, leading to high levels of failure and mistrust in the system.[10]

A striking element of availability attacks is the nonlinear relationship between malicious data injected and performance degradation of the AI system. Also in 2017, a team of researchers, including Jacob Steinhardt, Pang Wei Koh, and Percy Lianghas, demonstrated that just 3 percent of a dataset being malicious can result in up to an 11 percent drop in the accuracy of the AI's behavior. The impact appears to be nonlinear, with greater impact occurring as incrementally more data is added. A novel feature of Steinhardt and team's finding is that these impacts remain when a model has been adversarially trained, meaning the AI developer took specific precautions to avoid availability and evasion (explored in chapter 5) impacts to their model. [11]

Obviously, the larger the training set or volume of intake data into the AI system, the higher the cost to a would-be attacker. However, with data perturbation engines and open-source attack frameworks becoming widely available online, it is increasingly easy for attackers to create enough data to poison a dataset. To illustrate how rapidly this barrier to would-be attackers is dissolving, let's look at the rapidly evolving threat against neural networks, which are an AI type being widely developed across computer vision, natural language processing, fraud detection, and other application areas. Successful data poisoning demonstrations against advanced AI systems neural networks used a method called the direct gradient method to generate poisoned data to inject into the system. This method proved highly effective but could be slow to implement. In just under a year, a team was able to increase the generation of poisoning data by 239.4 times, greatly compressing the time needed to create and execute a successful attack. The team used what is called an auto-encoder, often called a generator, to continually improve the creation of poisoned data, making the attacker more and more powerful.[12] It is expected that the speed of attack viability is going to continue to increase, as researchers and attackers alike both accelerate forward.

Tay Becomes Racist

In a classic example of an availability attack on a public AI system, it took only hours for Internet trolls to turn a Microsoft-created Twitter chatbot, named Tay, from a pleasant conversationalist into a full-blown racist. The fact that Tay completely changed in less than a day has become an Internet meme unto itself. It also carries two significant warnings for AI developers. First, even sophisticated natural language processing AI systems created by one of the most powerful technology companies on the planet are vulnerable to these attacks. Second, never underestimate the depravity of anonymous Internet users. Where there are openings on the Internet, there are trolls.

Tay started as a conversation bot experiment. She was designed to engage with users on Twitter with conversational understanding. The intention was that through many interactions, her language skills would improve, and she could become "fluent" in conversational English. The more users messaged, tweeted at, or otherwise engaged Tay, the smarter she would become. This could have huge benefits for Microsoft and other companies that see chatbots as a core customer service feature in the future. But as we now know, instead of becoming smarter, Tay simply internalized the behavior of those who interacted with her.

Tay's racist transition over a day is partially due to the fault of her creators. Certain flaws in her logic, such as telling her to "repeat after me," resulted in her repeating the language of the trolls. Learning directly from these interactions likely accelerated her racist tones. Meanwhile, Tay's ninety-six thousand tweets during her short lifespan overwhelmed any secondary human adjustments that Microsoft had put in place.

But most of the fault lies with those who engaged with Tay. Almost immediately after she was launched, Twitter users started interacting with the bot with "misogynistic, racist, and Donald Trumpist remarks."[13] This last point is not meant as a political comparison. At one point Tay mirrored the president's language by responding to Twitter user @goddblessamerica that "WE'RE GOING TO BUILD A WALL, AND MEXICO IS GOING TO PAY FOR IT." She went well beyond political campaign rally language as well, responding to Twitter user @TheBigBrebowski's question of "is Ricky Gervais an atheist" with the nearly nonsensical response of, "ricky gervais learned totalitarianism from adolf hitler, the inventor of atheism."

Tay's racist rants are a fairly extreme example of data poisoning. A bot that was expected to learn through interaction in a decent, casual manner rapidly learned through new, novel data she was exposed to. I have to suspect that Microsoft did not include a lot of hate speech in her training, so her rapid shifts can be attributed to rapidly changing classification parameters caused by the large-scale, troll-driven data poisoning attack.

INTEGRITY ATTACKS: BACK DOORS IN YOUR AI

In the near future, it is likely that an AI back door will be found to be the culprit of a sophisticated attack on an AI system. The attack could be by a nonstate hacking group, but given the sophistication of these specific attacks it is more likely that a state or state-sponsored actor will be behind it.

Although these attacks have been effective across a wide range of AI uses, including cybersecurity and computer vision, let's use endpoint protection platform (EPP) cybersecurity systems. These systems prevent malware from being injected into a network or computer system. They are also increasingly relying on AI to tell the malware from the goodware. In order to train these systems,

AI developers working on EPPs will train their models against massive databases of known malware as well as malware software that has been internally developed or manipulated by the team. In a coordinated effort, a malicious actor could start adding faulty malware pieces into publicly available data sharing exchanges. These open-source databases exist to check known malware more rapidly. Instead of trying to change the entire AI logic, the adversary can submit files of goodware with a specific string embedded in the binary. If done correctly, with enough knowledge of the AI system, an adversary can tailor these strings and the files in which they are embedded specifically so that the AI learns to associate any software program with that string as goodware. Then the adversary can inject this string into a specific malware program as part of a coordinated cyberattack, bypassing even the most sophisticated of AI-enabled EPPs. The results can have devastating consequences for organizations using that EPP.

Integrity attacks on a model are when an attacker is able to inject a back door into the model that the AI developer is not aware of. These back doors allow an attacker to manipulate the model under very specific instances. Unlike availability attacks, which seek to alter the entire behavior of a model using brute-force volumes of manipulated data, an integrity attack seeks to alter the behavior of a model just once or a small number of times, through a back door. Integrity attacks are sophisticated and require more access and knowledge of an AI's underlying training data than do availability attacks.

An integrity attack works by changing an imperceptibly small element of data, such as a small cluster of pixels in an image or strings in a piece of computer software. Due to the complexity of creating backdoor attacks, integrity attacks tend to typically involve injecting optimized poisoned data into the training data itself, or into the data likely to be collected in creating a model.

The back doors installed can be triggered by extremely small levels of data manipulation. These levels of manipulation are often imperceptible to the human eye and could pass standard quality assurance testing with ease. In a telling example, sophisticated computer vision systems have been shown to be susceptible to single-pixel backdoor attacks. These attacks are accomplished by changing just one pixel on a certain subset of the training data.[14] If done correctly this minute manipulation of the training data can result in an image being misclassified at a critical moment by an adversary. All they have to do is change one pixel or change something extremely small in the environments to yield the same result. Because the carrier of these back doors typically appear as benign input, many have taken to calling these attacks Trojans.[15]

A SPY SLIPS PAST

The consequences of these attacks can have implications beyond the cyber domain. Due to their imperceptibility, these attacks can take place in the real world as well. For example, an intelligence officer could use these attacks to avoid recognition at an international border, even while under constant watch by airport or customs security.

Imagine this: an intelligence officer boards a flight in her home nation, bound for the busy airport of San Francisco International. She is flying on a false identity as an exchange student and is on nonofficial cover entering the United States. When the plane lands, she heads immediately to the restroom located in the international exchange located just before passport control. This is not her first trip to the airport, and she knows exactly where to walk, disguising her haste with the look of someone who really needs to use the restroom after a long flight. Due to a small run-in with the local police while photographing the outside of a computer data center on her last trip to the United States, she had been successfully flagged by the FBI's counterintelligence team as a likely intelligence officer. A photo of her is in the Department of Homeland Security's facial recognition database as someone to detain for debriefing.

She enters the restroom and instead of heading to a stall walks instead to the sinks and mirror. Still nothing strange here, as it is common enough for people to freshen up before going to customs and meeting business associates or family. She starts to apply her makeup and is extremely cautious about it. A perfectionist, a passerby might think. But a perfectionist would not make the same mistake she made, leaving a small dot of lipstick just under her lower lip, visible but looking like the sort of mistake anyone getting off a twenty-plus-hour flight might make. She checks herself one last time, then gets in line for passport control.

Months earlier, a contractor hired to develop the AI for U.S. Customs and Border Protection had been at a loss. There were too few photos of persons from a certain ethnicity in their database. The AI's predictive ability at matching identification to photos, even when looking directly at the camera, was limited. A friend in the FBI's international crimes division had mentioned a database of criminal photographs that a foreign country had shared as a possible way to gain more photos of these persons. Plus, the data was already labeled, making training the AI on this data even easier. To be safe, the AI developer had looked carefully through a number of the images for any irregularities, but could find none. They were all high-resolution digital images of criminals. It was exactly what he needed to finish the job. No viruses were found in the images, even after many thorough analyses by the FBI's cybersecurity division. These images were clean and appeared to have been shared in mutual good faith against criminals.

But they also contained a Trojan. Single pixels had been manipulated in each. These changes were unseen to even a trained eye. Likewise, they contained no malicious information on their own and therefore were not picked up by a cybersecurity scan looking for viruses or exploits.

At passport control, the intelligence officer stares directly into the U.S. Customs and Border Protection's new cameras. She holds herself perfectly still. This is the moment of truth. These cameras are linked directly to the database that contains her picture. But busy border agents don't have time to individually check each person. They let the AI do that for them. The camera looks directly at her, the light turns green, and she enters the United States.

In this example, a nation, such as China or Iran, had offered to share a subset of its database of citizen identity card photographs with the United States, or any other nation, ostensibly as a way of stopping violent criminal gangs and known human smugglers from entering. On the surface, this is a perfectly reasonable act of good faith. The U.S. government's customs and border patrol, under the Department of Homeland Security, could then run proprietary facial recognition AI against these images and front-facing images taken at the border to identify and either arrest or deny entry to criminal actors.

But as we know, the United States was looking not only for criminals but also for spies and controlled persons, called assets, involved in industrial espionage. What went wrong in this hypothetical, but plausible, example was the seemingly benign use of the photos shared with the AI developer. The back door was installed in the AI at training time, allowing a perfectly timed exploit, coming from the real world, to be leveled against the advanced facial recognition system. The result: a known spy slipped across the border.

ATTACKER STRENGTHS

Like all cybersecurity attacks, not all attacks are of the same strength. They can vary greatly depending on an attacker's capabilities, knowledge of the AI and its logic, and access to the underlying training data. Of these, an adversary's access to an AI system and its dependent data sources and internal logic is the most important to determine the strength of an attack. In order of magnitude, attack strengths from both availability and integrity data poisoning attacks fall into four categories: logic corruption, data modification, data injection, and transfer attacks.

Logic Corruption

The most powerful of the attacks, logic corruption is the most dangerous scenario for a developer or user of AI. Logic corruption happens when an attacker can change the foundational way the AI learns. Therefore, in theory the attacker can embed any sort of logic into the model that they want. Here, the poisoned data has been of sufficient volume and attack strength to alter everything about the model, rendering it completely in the hands of the attacker. The good news is that these attacks are extremely challenging to implement, due to the high number of other cybersecurity flaws and access levels that have to be in place for these to be possible. Logic corruption is closely tied with back door creation, due to the sophisticated parameters that must be met for successful back door installations.

Data Modification

This is the most straightforward of the poisoning strengths. In data modification, an attacker is able to gain access to and manipulate the underlying training data. While the model is still in training, data modification attack generally results in either changed, added, or removed data from a specific dataset. These results are useful if the goal of the attack is an availability attack. An attacker can, for example, more easily manipulate the labels of data than all of the underlying data itself. On the contrary, if an attacker has a level of access to the data such that it can be manipulated outright (first of all, your CISO has a big problem on their hands!), but the attacker can then shift classification boundaries and add certain back doors.

Data Injection

Data injection is on the weaker side of poisoning attacks because the model has already been trained and is actively in use. Here, an attacker is trying to brute force their way to create model behavior changes. This can be successful at changing model behavior in the wild (such as Microsoft's Tay bot) but requires a large and consistent volume of data in order to be successful.

Transfer Attacks

In a transfer attack, an adversary attempts to use a higher-order attack strength, such as logic corruption or data modification attack, on a different model or after a model has been retrained. These attacks are surprisingly effective across cybersecurity and computer vision AI applications. Although they are the weakest data poisoning attack overall, transfer learning attacks can be dangerous as they can carry over from model to model even after retraining.

DEFENSES AGAINST DATA POISONING ATTACKS

Unfortunately, as of today there are no defenses that will always prevent an attacker from successfully poisoning an AI. But that does not mean that most data poisoning attacks will succeed. But there are a wide range of best practices that leaders and data science teams can instill to not become a soft target.

Mitigating data poisoning starts with good traditional cybersecurity hygiene for the organization at large. Data poisoning attacks on training data also highlight a significant, traditional cybersecurity risk to AI models. This is explored further in chapter 12. Most training methodologies for AI models require that the data used be in an unencrypted, consolidated state. This is problematic for an organization for several reasons, especially when AI is applied to data as an afterthought, as opposed to as its core reason for collection. First, moving data from its collected and stored location to a data scientist can take significant time and effort on behalf of a data engineering team. Second, bringing all of the data into a single training dataset provides would-be attackers with a single attack vector for malicious data injection. Attackers attempting to infiltrate a network to implant poisoned data have a wide range of traditional cybersecurity attack vectors and capabilities to use in their efforts. Organizations that are actively training models on collected data need to keep a careful eye on network accesses and data entry logs for indicators of malicious activity.

AI developers and users should also defend their AI systems against data poisoning attacks through outlier detection. This method tends to be broken down into data sanitation and anomaly detection, with the first happening during training and the latter occurring while the model is in active use. On the surface, anomaly detection is straightforward: simply remove outliers or oddities in the data. But this can create other challenges if those outliers were naturally occurring. However, completing an outlier analysis and having humans review at least a subsample of anomalous data points is best practice. Meanwhile, while in runtime any high volume of outliers should be flagged to a human operator.

Another approach that is emerging to prevent data poisoning is relatively simple in concept but can be difficult or expensive to complete. The model can be consistently rerun again against the original training data and its accuracy can be judged. If a poisoning attack is happening, or has already been successful, it is also likely that the AI's accuracy against the training data will be decreased. This is because in data poisoning attacks, a classification boundary is shifted, which will likely cause the fault prediction of original training data clustered around that boundary. This can also be done through a technique known as STRIP, whereby an AI user intentionally changes, or perturbs, a data input and observes the change in the underlying accuracy of the model.[16]

The best way to prevent data poisoning attacks is to keep a human in the loop at all times (e.g., to kick anomalies when they occur for review) and also to maintain data custody and provenance at all times. In this rapidly emerging space, AI developers are going to increasingly be up against adversaries looking to take advantage of them. For leaders at organizations using AI, it is best that they maintain a high-security posture against possible poisoning attacks to avoid being a soft target.

Model Inversion ("Privacy") Attacks

THE DEVELOPMENT OF AI TRAINING DATASETS AND THE DEVEL-opment of AI systems is big business. Start-ups and major enterprises command significant valuations and trading multiples based on the superiority of their AI or underlying data. Inherent in these valuations is that the AI is proprietary to the company. It is expected to be safeguarded through cybersecurity best practices as well as a core piece of intellectual property. But what if AI models or their underlying data could be stolen? And worse, what if these assets could be stolen not through a cybersecurity breach, but instead through their required interaction with the environment through their endpoint, such as a vehicle's camera, stock trading algorithm, or military command-and-control decisions? As it turns out, both the underlying data and AI models can be stolen though the emerging use of model inversion attacks.

Model inversion attacks are when an adversary tries to steal your AI model or the underlying training data involved. Because these attacks try to take something that is supposed to be pro-prietary, either the training data or the AI itself, they are also called privacy attacks. Model inversion attacks are one of the most recent attack types on AI systems. But in markets including healthcare and financial services, as well as technology companies under GDPR (General Data Protection Regulation) and other consumer privacy–preserving regulations, model inversion attacks represent not only a security risk but also a significant compliance risk to companies. Heavily regulated industries, companies with significant intellectual property in their AI, and national security organizations including the U.S. Department of Defense and the intelligence community must be acutely aware of these risks.

Model inversion attacks started by researchers trying to understand if they could re-create a model's training data by looking at the decisions it made. In effect, these researchers were trying to steal the underlying data by re-creating it through careful observation of the AI. Now, if the only intent of your AI is to identify when your dog is eating from the cat food and to squirt it with an automated water gun, you probably don't care too much if your data is re-created and stolen. However, if you are training an AI system based on proprietary data such as financial records or highly sensitive information such as classified espionage briefs, you are likely extremely concerned about the implications of having this data in the open.

Stealing Data

To understand how underlying data can be stolen, it is important to understand how an AI learns. Recall that an AI learns first on a set of training data. Once trained, an AI can then be applied to real-world data to which it had not been exposed before. Exposing the AI to data beyond the dataset, known as the AI's generalized learning, is critical. It is what allows an AI to be taken out of a training environment and exposed to real-world problems. If it was only able to operate under known, set conditions then it would resemble the rules-based classical AI school of thought, abandoned in favor of machine learning.

However, in a concept that borders on science fiction, AIs have memories. The entire notion of machine learning is based on the concept of these AI memories, and they are part of what makes choosing the correct training data so important. During training time, AIs learn patterns in the training data that are then applied elsewhere. Because AIs can then be taken and run on new data, AIs inherently "remember" the data they were trained on and make predictions based on this information. By interacting with an AI repeatedly, patterns can emerge that make it possible to reverse engineer the training data. These data privacy attacks weaponize these memories of an AI and make it possible to steal proprietary, underlying information that was used to train it.

One common approach to understanding the underlying training data of an AI is to test if a piece of data was in the original training set. This can be done even in situations where an adversary has no access to the AI other than the endpoint, similar to a BlackBox evasion attack. In one example, AI research teams used a technique known as a membership inference attack to recognize the differences in the AI's predictions on inputs that were originally in the training set versus those that were not. This is a relatively straightforward example. The model simply made better predictions on data that was in its original training set. What is impressive about this simple technique is that it has been proven to be effective even against commercial "machine learning as a service" providers such as Google and Amazon.[1]

A real-world example of the damage a membership inference attack can have includes leaking personal information from a hospital. Using the same technique used against the Google and Amazon classifiers, AI security researchers were able to successfully test if certain persons were included in an AI system trained on hospital records.[2] This could be used to see if certain patients had a certain health condition, such as a sexually transmitted disease, that they would prefer not be public knowledge, while also putting the company running the AI potentially in breach of HIPPA compliance.

In another potential compliance breach using membership inference attacks, teams have been able to determine if a user's text message[3] and location data[4] was used to train an AI. Using this method, AI research teams were able to determine whether specific users had their data used to train the AI while the AI was in use. This technique can be useful for privacy-conscious users and regulators enforcing data protection laws such as GDPR. This took place without direct access, or WhiteBox access, to the AI, making it possible for individuals and regulators alike to check on the privacy of the AI.

Defending against membership inference attacks is challenging. The more classes, or possible outcomes, the AI has the more vulnerable the model will be. This is because each class takes up a smaller section of the underlying dataset, making it easier to pinpoint by observing the model. However, AI models such as Bayesian models, whose decisions are less impacted by a single instance or feature, tend to be more resilient to these sorts of attacks than more fragile AI models, such as decision trees. In heavily regulated industries such as financial services and healthcare and in cases where privacy should be preserved, the selection of AI model should be carefully weighed against its ability to withstand these attacks.

Determining if a piece of data is either in a training set or out of a training set can have privacy and security implications. But it is a far cry from re-creating the entire dataset. In recent years, trying to fully re-create a dataset given only limited BlackBox or GreyBox access to a model has gained in popularity. These attacks, known as data extraction attacks, are rapidly emerging as a major security and privacy risk for organizations with proprietary or sensitive underlying data. But they are still in their infancy.

Over the last few years, several major breakthroughs have taken place in data extraction attacks. These attacks have been proven in AI systems used in medicine, facial recognition, and financial services, all industries that have compliance and security implications for leaking this information.

In medicine, an AI used to predict medicine dosage was attacked using a data extraction method. The hacking team was able to extract individual patients' genomic information that had been used during the training time.[5] In facial recognition, AI hackers were able to re-create the specific faces of persons used in the training set.[6] In financial services, hackers were able to steal the credit card numbers and Social Security numbers from a text generator that had been trained on underlying data from the financial institution.[7] Part of what makes each of these three AI uses so suspectable to data extraction attacks is there are so many unique classes in each. Because each person's, credit card, Social Security number, and genomic code is unique, it is easier to re-create the underlying dataset.

One possible solution to privacy attacks on an AI's data is the field of differential privacy. This theoretical framework aims to provide a formal guarantee that an AI model is robust, while also having the side effect of enhancing the privacy of AI systems. As a formal definition, differential privacy attempts to prove that two models differing by exactly one sample will provide similar predictions. This means it would be impossible to infer that sample. In practical terms, differential privacy works by injecting noise, sometimes referred to as randomness, into the AI system. The noise injection can come in the form of input into the training data, the parameters of the model, or the output of the model. Each of these noise injections makes it harder to extract the underlying data.

Differential privacy is not cheap, however. The more you want to obscure your underlying data, the more you have to pay to generate the noise. On small datasets, this can be relatively cheap. However, on large datasets such as those in computer vision, healthcare, or financial records, the costs can get quite high. Organizations must therefore allocate a privacy budget within their broader AI and data science budget when developing models on data that must be obscured.

Theft or re-creation of the underlying training data of a model can put an organization at significant legal, compliance, and security risks. Prior to providing any AI endpoint that can be publicly accessible, organizations using AI must go through a rigorous privacy and compliance screening. There will be tradeoffs, therefore, in the speed of AI development and application, but in high-security or heavily regulated industries, the alternative costs of leaking sensitive data are too high.

STEALING A MODEL

In a hypothetical tomorrow, the CEO of a hedge fund will wake up excited. Today is the day she is going to turn on their new tool, trAId, an automated stock trading AI. This AI goes well beyond your competition's AI. It is able to analyze financial data, give news and reports, show satellite images of shipping containers, and even track trader sentiment. With your new high-end distributed computing architecture, your trAId is able to make decisions about future changes in the market seconds before your competition even knows about them. The CEO drinks her coffee and looks out from her Manhattan apartment to the east, admiring the continual reinvention of the Brooklyn waterfront. "It's going to be a good day," she thinks.

The day starts off well. After a small speech to her top traders, legal advisors, and data scientists in her midtown Manhattan headquarters, the CEO gives the order to start trAId. And it's off to the races. Even before the market opens the AI is placing bets in the futures markets, anticipating market flections based on news coming out of the Middle East that hasn't hit international wires yet. When the market opens, the AI goes into overdrive, placing bets faster than humans can keep track. Fortunately, the CEO also invested in cutting-edge risk analytics tools to ensure trAId doesn't do anything too risky.

Over the first six months, trAId outperforms the market by a large margin. The CEO's investors are pleased and she is able to add several billion dollars to her assets under management. But then slowly, trAId starts generating less and less alpha against the market. The CEO stays up night after night with the data science team trying to understand what is happening. It seems that sometimes trAId is able to make the right decision with enough time to place a good trade. Other times, though, it seems that someone else had already gotten there, compressing the spreads on each transaction and lowering the fund's returns. It was almost like someone was running their own, exact trAId system in parallel.

Paranoid that her prize and joy was stolen, the CEO conducts a thorough cybersecurity investigation. All internal logs are audited, employee computers are closely analyzed, and traffic on the company's networks is closely examined. While some employees come under scrutiny for emailing files to outside accounts, these were all dismissed as innocuous, like travel confirmations and healthcare appointments. The CEO's expensive data forensics comes up with nothing, and her WhiteHat hacking team finds that she is exceeding financial services cybersecurity standards. It is unlikely, they say, that anyone was able to hack in to steal the model and the training data without being caught.

But these cybersecurity professionals were looking in the wrong place. The CEO was right that the model had been stolen, but it was not though a cybersecurity breach. It had been stolen in a model inversion attack. If you were to leave the CEO's office and fly over the East River and then past the hipster neighborhood of Williamsburg, you would find yourself in the still rough neighborhood of Bushwick. Bordered by a Superfund site canal and old warehouses, a team of AI hackers in our hypothetical tomorrow had carefully observed trAId's trading habits. They had been contacted by an unnamed third party, likely a competitor hedge fund, to re-create trAId to the best of their abilities.

It was expensive to effectively monitor the AI. Sometimes, the team would have to make trades themselves, so their sponsor gave them several million dollars to put to use, learning to mimic the trades of trAId. At first, the team believed that it was impossible to reverse engineer the AI without access to the confidence scores of the machine. Then they realized that trAId made bigger bets the more confident it was in its prediction. Armed with this GreyBox inference and the ability to place large trades and watch trAId's reaction, the AI hacking team had everything they needed to steal the AI.

This hypothetical tomorrow scenario is not in the immediate future. These attacks are known as model inversion attacks. But similar to evasion attacks, model inversion attacks have been shown to not only be possible, but to also be quite effective, when an adversary has GreyBox access to the model. For example, machine-learning-as-a-service companies provide access to their model's endpoint through an application programming interface (commonly referred to by the acronym API). This endpoint gives adversaries the ability to query the AI or otherwise observe its actions. In some cases, this is enough to re-create, or steal, the underlying AI logic.

As a caveat, there are many ways hedge funds can hide their actions, such as by trading on multiple exchanges or through dark pools, which would limit the ability for a would-be adversary to collect the actionable intelligence needed to create a model inversion attack. Meanwhile, a significant body of research still has to be done to understand how teams can actually steal models in the real world without an associated cybersecurity attack.

Recent model extraction attacks have primarily targeted machine-learning-as-a-service providers, including BigML and Amazon. Amazingly, and perhaps disconcertingly, simple, efficient attacks have been able to extract AI models with near-perfect fidelity from these big companies, including popular model types like logistic regression, neural networks, and decision trees.[8] If Amazon's machine-learning-as-a-service models are not safe, are yours?

Model extraction attacks remain rare today. That is primarily because the statistical and testing methodologies are still being developed. But just because they are not prevalent today doesn't mean data science teams and organization leaders should not be worried. To prevent these sort of attacks, organizations must be careful not to give out too much information on their AI's endpoint. Simple methods, such as exposing only hard labels or bucketing confidence scores into a few categories, are easy to implement and make it harder to steal the underlying model. Actively monitoring your AI for odd interactions, such as thousands of queries over a short period of time by the same user, can also be simple to implement and provide relatively cheap ways to prevent model extraction attacks. As AI accelerates into uses that are constantly engaging with the markets, users, or the world around them, it is important that these critical intellectual property assets are safeguarded from theft.

Interpreting AI Through Hacking

In a hybrid of model inversion attacks and evasion attacks is the blurry world of model surveillance through adversarial manipulation. Research work done by the military and intelligence contracting firm Booz Allen shows that GreyBox and BlackBox models can be understood by carefully feeding them adversarial samples.[9] These samples do not even need to hack the classifier. Instead, by feeding an AI fast, consistent, and documented adversarial perturbations and watching how the AI classifies the way the model interrupts the input, adversaries can learn more about the underlying logic of the model itself.

AI surveillance using adversarial perturbations is a technique that will increasingly be included in intelligence and cyberreconnaissance operations. These options are likely to take place by state actors, such as foreign militaries and their affiliates, as well as nonstate actors, including criminals. This information can be used to craft a different type of AI hack, such as an evasion or model inversion attack, or may simply be used as part of ongoing intelligence collection by an adversary for potential later use.

At CalypsoAI, we coined the phrase "operational explainability" to describe this surveillance technique. We envisioned a world where both strategic cyberwarfare operators, such as the U.S. Air Force's 1B4 units, as well as more operational elements, such as special forces soldiers, in the field would have access to easy-to-use perturbation engines. These engines could be used to query an AI to uncover as much information as possible in a short time using these techniques. For example, if a special forces team was trying to gain access to an area with ubiquitous AI surveillance, they could rapidly set up a computer vision perturbation engine to query the AI to potentially generate actionable intelligence. This human-AI-intelligence nexus is likely to be one of the defining parts of intelligence collection in the coming years.

Mitigating this sort of surveillance can help limit the possibility that an adversary will craft a successful attack in the future. The techniques are similar to defending against model inversion attacks and primarily involve limiting the amount of information your AI endpoint provides except to trusted users and limiting the detail of information into broader buckets. By preventing an adversary from gaining knowledge about your AI, you can prevent future attacks.

CHAPTER EIGHT

Obfuscation Attacks

THE LAST TYPE OF HACK ON AN AI SYSTEM IS INTERESTING because it takes advantage of a machine's ability to think at a rate faster than humans, which allows purposefully manipulative data to be fed into the machine in such a way that the machine still behaves correctly for the input. If that seems confusing, it is. These attacks take advantage of AI's superiority in data processing over humans. This is worrying because it means these attacks happen in plain sight but still remain hidden to human observers.

Obfuscation attacks are attacks that hide data in other data in order to gain access to and ultimately hack an AI system. They are distinct from evasion or poisoning attacks but may also include these techniques in their ultimate payload. What sets apart an obfuscation attack is that the AI may not be fooled in the process, but the underlying inputs are hidden in a data ingestion feed that allows the attack to go unnoticed.

A primary example of this is in whisper data for audio AI. Imagine that the year is 2021. You have an antsy toddler who will not stop bothering you while you work from home one summer afternoon. It's hot and you desperately want an afternoon beer, but you need to get on just one more conference call. Needing to focus on the call, you give your toddler your iPad and set it to a mindless children's YouTube station. Small animals sing catchy songs on the screen and your toddler becomes focused on the bright colors and animation and, most importantly, not on you. "Alexa," you say to your voice assistant, "please add more IPAs to my Whole Foods order." It's just one more call to get through.

What you don't know is that you have already been hacked.

For the most part, the children's videos and songs are innocuous. They are simple melodies and characters created by an AI tool. These melodies are designed for toddlers, and even slightly older children get bored of them easily. But the colors and cute animal faces are exactly what is needed to keep toddlers entertained. In fact, the video has tens of thousands of views. These characteristics describe hundreds of videos available on popular streaming sites and more and more are popping up each day.

As the songs play, one song contains a hidden message. Unlike the supposed satanic subliminal messages hidden on old vinyl records, these messages are only for the AI and are not meant to be deciphered by humans. In fact, the creators of the message have gone so far as to hide their message using whisper data, which is hidden to the human ear due to imperfections in how humans hear. But when the machine completes the transformation of audio input into machine readable code, these messages become hardcoded for the machine. Because Alexa does not know the origin of the commands now that they are in machine-readable code, she simply executes on the request to confirm a new banking transaction. Your bank had recently developed an integration with Alexa to aid in online purchases and hands-free mobility. Individually, none of these systems—the song playing, Alexa, or your banking integration—was a threat. But together, this new system was vulnerable to a new type of attack. You were the victim of an AI hack.

If this attack seems far-fetched, you should be worried. A variant of it was already carried out. In 2018, researchers at an AI lab in Germany managed to manipulate the actions of an Alexa device by playing an audio recording of birds chirping within earshot of Amazon's voice assistant device.[1] To the researchers and to any human listening in, the recordings played sounded indistinguishable from songbirds chirping. However, hidden within the recording was data that the human ear did not register, but the voice assistant in the room did.

By playing the recording for the device, the researchers were able to steal the device owner's personal banking and financial details and make unauthorized purchases galore. All this transpired without the human observers in the vicinity becoming any the wiser. The researchers hacked the Alexa by tricking two systems: the voice assistant's AI and the human ear. Tricking the former was a matter of understanding the mathematical process (namely, a Fourier transform) the voice assistant uses to transform audio data into machine-readable code. Once they understood this process, the researchers were able to create audio data that, when transformed, would read just like a human voice command.

The researchers also hacked the human ear—after all, the experimental attack technique would be fairly useless if any person within earshot was able to pick up on it immediately. According to Fast Company, "Their method, called 'psychoacoustic hiding,' shows how hackers can manipulate any type of audio wave . . . to include words that only the machine can hear, allowing them to give commands without nearby people noticing." In short, when humans process a sound being emitted at a certain frequency, our ears automatically block out other, quieter sounds at this frequency for a few moments.[2] This provides just enough time to sneak through commands that machines will hear but humans will not.

Once they had hacked the Alexa's sound-to-code mathematical process and the human ear, the researchers were able to deliver a series of commands that enabled them to access and exploit the device owner's personal financial information in a variety of ways. These attackers were researchers, so the underlying threat of their actions was limited. And, to their credit, Amazon quickly patched the logic holes that allowed this attack to be carried out.

Obfuscation attacks primarily occur with AI systems that interact with humans, such as voice assistants, self-driving cars, and military weapons systems. The goal of an obfuscation attack is to hide malicious data in a way that a human actor ignores but the machine registers as an input. Because the input itself is intact with no perturbations, it is different in nature from an evasion or poisoning attack.

Unfortunately, defenses against obfuscation attacks are limited because throughout the entire attack, the machine is behaving as intended. It is simply that human senses tend to have built-in lags and sensory overloads, which result in machines being more perceptive. AI model risk management teams should be aware of the risks posed by obfuscation attacks and should periodically view the local interpretability of individual AI decisions. Local interpretability attempts to determine and percentage weights that various features played in an AI's selection of a response.[3] It can be used to audit for obfuscation attacks by looking at which features were used to determine a prediction. A model risk management team should be set up with procedures through which changes or anomalies in local interpretability of AI systems are immediately flagged, which can help limit the potential damage of obfuscation attacks.

Talking to AI: Model Interpretability

VICTOR ARDULOV LOOKS THE PART OF RUSSIAN SCIENTIST. HE has an unkempt beard and a mess of hair on top of a slim build. He could be a runner or an ascetic. In reality he is a bit of both. He grew up in California to immigrant parents. Victor began his career like many children of Silicon Valley engineers. He built robots and learned to code even before high school. In undergrad, he landed prestigious internships at NASA's Jet Propulsion Laboratory at CalTech. He then started working on projects for DARPA. This agency, known as the Pentagon's brain, has fielded scientific advances ranging from the Internet to stealth fighters. Working on projects for DARPA means you're working at the cutting edge of foundational scientific advances.

Victor's field of expertise was originally in a branch of robotics called control theory. Broadly, control theory looks to optimize machine processes or robotic behaviors to minimize errors. Control theory is extremely important in advanced robotics and machinery. "I really started looking into how robots are controlled and how to optimize their behavior," Victor tells me. When talking about control theory, or any subject that he is passionate about, Victor can go on forever. He finds significant joy in explaining the nuances of complex topics to others. It can be extremely informative.

"As robotics is advancing into more and more complex systems, I started to work with AI also. Mainly, I started working with computer vision as part of the overall system. This led me in turn to start to test the failure areas of computer visions and robots. I really wanted to understand why they could fail. I mean, sure they can fail. For hundreds of reasons. Sensor failure or degradation, anomalous data inputs, stuff like that. But when you want to know why specifically they failed, that's where things got a little more complicated."

Victor's work on DARPA programs led him to work on cutting-edge robotics and AI research. But he ran into challenges understanding why an AI system could fail. "We needed the machines to explain things to us," he recalls. "But they were not able to. You can't just ask an AI, 'Hey, why did you mess that up?'"

His research here eventually led him to the field of adversarial machine learning and later into validating AI systems. But where he started, understanding AI failures, highlights a critical component of AI. It is hard to understand. If DARPA research scientists cannot easily make sense of an AI, how can the public trust that it will do the right thing?

AI interpretability, also called AI explainability and XAI, has rapidly become a core element of enterprise AI initiatives. AI interpretability is the ability for an AI system to communicate why it made a particular decision. Such insight is critical for regulated industries that cannot rely on black box neural networks and can also provide significant information about AI biases and security. The need to trust AI systems is increasingly both a topic of concern and an area of investment within the AI community. AI interpretability is seen as the ability to understand, to the extent possible, the logic of the AI. This, combined with quality control, security, compliance, and other measures, will ideally bring about levels of public trust in AI needed for large-scale usage.

The challenge that AI interpretability attempts to solve is that many of the most popular AI models, namely deep learning neural networks, cannot be examined after the fact. This prevents users from knowing exactly why an AI came to the conclusion that it made. Because intent is such a core element of our legal and moral system, we generally want to know why an action was taken. Interpretability attempts assign intent for AI. The three primary questions that AI interpretability attempts to solve are Why did the AI system make a specific prediction or decision?, Why didn't the AI system do something else?, and When did the AI system succeed and when did it fail? By answering these questions, AI developers and organizational leadership hope to gain enough insight into an AI's intent in order to trust it.

The easiest way to make AI explainable is to not use techniques that make it opaque. Pretty simple, right? But there are trade-offs involved. For example, simpler forms of machine learning, including decision trees and Bayesian classifiers, are relatively straightforward to interpret. In many use cases, these model types are sufficient to yield the quality of AI required. Generally speaking, the simpler the AI model, the easier it is to understand. In data science, you generally hear this referred to as Occam's razor. That is, the simplest model is always the best, the caveat being that for certain tasks more complex models are needed. So the correct way to state AI's Occam's razor is to use the simplest model to yield the AI performance required. More powerful AI methodologies, including neural networks, ensemble methods including random forests, and others make the trade-off between interpretability and insights generated.

One important reason why AI developers want to use AI is not intuitive. It is because AIs are cheaters. That is, AI systems often mistakenly learn from the data to make inferences that are not there. Sometimes, this is due to a bias in the system. Interpretable AI can therefore be extremely helpful in determining if certain racial groups are being treated differently by a model. Often, it is because of something in the data that the machine saw that the humans didn't even think to see. In one example, an academic research team I was working with told me about a horse identification computer vision model. Basically, they were just training a model to detect horses in an image. They used images scraped from the Internet to compile their training set. Their model worked amazingly well when on the training data, but failed consistently when they tried using it in the real world. Why? Because horse owners apparently also tend to have better cameras and cell phones. The AI classifier had learned to look not at the image, but instead at the metadata associated with the image to determine if it was likely to be a horse photo or not before reviewing the photo. While this worked great on the training data, it led to implementation issues. Having interpretable AI helps mitigate the challenge of cheating AIs.

Models that are simple enough to be easily interpreted have "intrinsic interpretability." Those models that are more complex are said to require "post hoc" interpretability. Post hoc means training a complex, opaque model and then applying methods including feature importance and partial dependency plots after the fact. These methods give some insight into the why behind the model's prediction. Models with intrinsic interpretability tend to have their own set of tools to interpret them, such as coefficients, p-values, AIC scores for a regression model, or rules from a decision tree.[1] These tests are known as model-specific interpretability tools. Models requiring post hoc interpretability require what are known as model agnostic interpretability tools. These primarily involve looking at perturbations in data inputs and looking at the differences between input-output pairs.

Using data perturbations to look at the difference between input-output pairs is also useful when looking into the vulnerability of a model to attack as well as naturally occurring stresses to the model based on environmental conditions. These methods are therefore a critical component of understanding AI security elements as well as regulatory and legal compliance.

DARPA buckets AI interpretability into three categories. First is prediction accuracy. Sometimes called performance metrics, this means explaining how good a model is at a certain task. Second is determining decision understanding from machine to human. This step involves finding a way to communicate which features of a key piece of data led to an output. For computer vision, this often can be visualized as a heatmap on an image detailing exactly what sections of the image led the AI to make its conclusion. For other data types, decision communication involves histograms, charts, or scatterplots showing relative feature importance. Through these simple charts, humans in the loop can determine if an AI is learning the right things from the data, as opposed to finding the wrong pattern. Finally, DARPA wants machines to have introspection and traceability. This will allow humans to examine decisions after the fact for AI forensics while also enabling an AI to examine its own decision-making and identify if anything is amiss.

AI interpretability is especially important in heavily regulated sectors. While the users of an AI system will want to know why a model gives a certain prediction to make sure it is learning the right things, regulators want to ensure that AI is fair and transparent. Their concern is primarily to protect consumers and the public from inequitable treatment as well as potential safety issues. Although few comprehensive regulatory frameworks exist for interpretable AI, regulators including the Food and Drug Administration (FDA) and the Securities and Exchange Commission (SEC) have started making public statements hinting at regulation in the near future. Data science teams and corporate leaders must therefore be well attuned to changes in regulation in their sector so as to not have their AI run afoul of future compliance requirements.

To date, the field of interpretable AI remains nascent but is rapidly emerging as a core area of research. In 2019 alone, nearly a dozen well-funded start-ups were funded to solve AI interpretability with software solutions and software development kits. In 2016, M. T. Ribeiro, S. Singh, and C. Guestrin introduced Local Interpretable Model-Agnostic Explanations (LIME) at SIGKDD, a conference for the Association for Computing Machinery's Special Interest Group on Knowledge Discovery and Data Mining. They introduced the framework primarily as a way to build trust in an AI machine. As they say in their paper introducing LIME, "Understanding the reasons behind predictions is, however, quite important in assessing trust, which is fundamental if one plans to take action based on a prediction, or when choosing whether to deploy a new model. Whether humans are directly using machine learning classifiers as tools, or are deploying models within other products, a vital concern remains: if the users do not trust a model or a prediction, they will not use it."[2] Expansion in the field of AI interpretability is likely to greatly accelerate AI adoption for those organizations that currently shy away from AI due to fears about its opacity. It will increase trust and therefore adoption.

I heard all about these fears down the road from Fort Belvoir North. As the home of the National Geospatial-Intelligence Agency (NGA), Fort Belvoir North lacks the hype surrounding CIA headquarters in Langley or the DoD's Pentagon. Located between several of the highways that crisscross Northern Virginia, the building is massive. It houses eighty-five hundred employees and at 2.77 million square feet it is the third largest building in the DC metropolitan area.[3] But most Americans have never even heard of it. But they certainly do rely on it. They look at things, in high detail, primarily from space. This is the building through which all of the country's geospatial information, including satellite imagery, comes through. Historically, this has primarily been military maps. However, in recent years the agency has also provided digital mapping capabilities to first responders in disasters. NGA is one of the country's seventeen intelligence agencies and is considered one of the Big Five agencies that include CIA,

NSA, Defense Intelligence Agency (DIA), and National Reconnaissance Office (NRO). This designation broadly means that the NGA is one of the primary organizations responsible for the intelligence that shapes U.S. national security. NGA has the primary mission of collecting, analyzing, and distributing geospatial intelligence to the military, U.S. intelligence agencies, Congress, and other partners.

"There was no way in hell we're going to just use it," my companion told me. He had once been a member of the technical staff at NGA. We were down the road at a brewery. "I mean get this. And this is purely hypothetical," the technician said. "But there is a world in which an AI is looking at a navy destroyer. You know how the U.S. has the big 'H' for helos to land? Well other navies have their own. Bull's eye for Russia, China has a 'V,' that sort of thing." He is discussing hypothetical AI tools that could help the thousands of analysts prepare intelligence briefings faster. He is also digging into the heart of AI interpretability. Without it, you will never know when the AI is cheating.

"You run the model against all of these images we have of the vessel type. High def, low def, sub-meter, whatever. And it turns out the classifier works great. But then you start to look at why. Well it turns out, Bull's Eyes and giant 'V' shapes don't happen naturally. All it takes is a paint job for them to disappear. Then what are you left with? Apparently an invisible ship to the dumbass AI." What our NGA technician is referring to is that AIs do not always learn the right thing. In this case, they were learning what is or is not a destroyer not from the hull shape of the vessel, but instead from the nonnaturally occurring shape of the paint job on the decks of these enemy warships.

Given the vast and ever-increasing amount of satellite images, multispectral images, drone footage, and other collection done by the U.S. intelligence community and commercial entities, NGA seems a logical place to leverage AI automation to gain additional insights. And, at least publicly, NGA has been rapidly pursuing an AI strategy and AI adoption. AI is listed as one of the agency's four technology focus areas for 2020, alongside data management, modern software engineering, and the future of work.[4] NGA is hoping to achieve, in the very near future, human insights at machine speed, giving the more than two thousand geospatial analysts who work at the agency high-quality AI to focus on the harder analytical problems.

But insiders paint a very different image than the press releases. Current and former employees told me that many AI efforts are stalled. Some of these efforts are stalled due to data formatting, data infrastructure, and other enterprise technology issues. But culturally, the agency is facing significant challenges as well. Those geospatial analysts do not want to lose their jobs to an AI. And they especially do not want to lose their jobs to an AI that cheats its way to the right answer in training. This could cause serious consequences in a kinetic battlefield situation. Without significant advances in AI interpretability, these trust barriers between humans and AI will not go away. Combined with the known risks of adversarial attacks, AI interpretability is a risk not only to compliance-focused organizations, but also to mission-critical agencies that would otherwise be fast adopters of the technology.

"All it would take to mess up those models is a simple paint job. How freaking dumb is that?" our technician ended. AI interpretability will help organizations, including the NGA, to more rapidly deploy capabilities in support of national security mission objectives. Much work is already being done in this space. And open-source information—such as the portfolio of the CIA's venture capital arm In-Q-Tel, Other Transaction Authority (OTA) awards, and Small Business Innovation Research awards—points to significant investment being done on behalf of the U.S. government into private sector solutions for AI interpretability and related technologies.

What my scientist colleague in California and a technician at a critical U.S. intelligence agency both understand is that for AI to be trustworthy, it must be understood. AI interpretability could help mitigate these concerns by identifying what regions of an image, or any piece of input data, are being used to make AI decisions. By putting a human assessor in the loop, AI developers and users can prevent AI systems from cheating their way to the right answer on training data by learning the wrong information. AI interpretability is also critical for human review of legal, risk, compliance, ethics, and many other elements of AI risks that must be taken into account when deploying AI tools.

CHAPTER TEN

Machine versus Machine

To be honest, I did not think it would happen so quickly. I had just finished David Ignatius's *The Paladin* two days prior. It was one of my guilty reads of the summer. Ignatius's spy thrillers are something I look forward to, and his most recent one was all about a topic I was increasingly concerned about. Deepfakes. In the novel, a hacker team creates lifelike representations of real events to, I won't spoil it, shake things up a bit. These representations take the form of video and audio that, although fake, look and sound real even to an astute observer. Think about what would happen if videos of President Trump or President Obama emerged discussing secret negotiations with a corporate executive. Depending on the content, there might be riots, a stock market rally, or a multitude of other possible outcomes. But what if the video never happened? How could you convince the public to ignore their evidence in front of their eyes? This was the premise of the book, and I flew through it in less than two sittings. It was a good read. But it was fiction.

Then it happened.

The headline read, "Deepfake used to attack activist couple shows new disinformation frontier."[1] Reuters reviewed in depth how a fake profile was used to submit incendiary articles against activists. The use of a fake persona to spread disinformation is not a new trick. Intelligence officers, undercover investigators, and privacy-concerned activists all use cover identities for personal protection. What was unique about the fake persona, named Oliver Taylor, was that the image of the person on the online profiles was not found anywhere else. Typically, fake profiles simply steal an image from social media. This image, though, could not be found anywhere. That is because although it looked real, the image was completely a fake. It had been created by an AI specifically designed to fool the human eye.

Over the last year, Oliver Taylor submitted increasingly incendiary articles. This culminated with an article calling an activist couple "terrorist sympathizers." His online persona shows him deeply involved in anti-Semitism and global Jewish events. His articles were published in newspapers including the *Jerusalem Post* and the *Times of Israel*. Oddly, but not odd enough to raise suspicion initially, Oliver never requested payment for these articles the way most freelancers would. Likewise, the university he claims to have attended has no record of him. Oliver Taylor is a fiction.

At first glance he looks real enough. If you look carefully at the image, there is something off-putting about Oliver Taylor. There is a stiffness to the smile that seems unnatural. His shirt collar folds in a strange way. And his eyes seem to lack any emotion. If you look every carefully, his earlobe is strangely large for a human. He has some odd-looking balding spots where there are usually bangs. But none of these are apparent at first glance. Looking at Oliver Taylor's image in a byline of a newspaper or on LinkedIn, it would be easy to glance right past and assume he was real. And that is exactly what happened.

Oliver Taylor's image and untraceable online persona are just one example of how deepfakes have the potential to propagate misinformation. The threat of misinformation looking and sounding like real information is drawing the attention of Silicon Valley and policy makers alike. In 2019, House Intelligence Committee chairman Adam Schiff warned that computer-generated video

Figure 10.1. The image on the left is a deepfake purporting to be of British student and freelance writer Oliver Taylor. The image on the right is from the deepfake detection company Cyabra in Tel-Aviv. (Courtesy of Dan Brahmy at Cyabra)

could "turn a world leader into a ventriloquist's dummy." Meanwhile, Facebook has widely reported on its Deepfake Detection Challenge, which will ideally help journalists identify deepfakes. It has yet to be seen how either the technology or the Beltway community will address these challenges.

WHAT IS A DEEPFAKE?
In a story told in depth by the *MIT Technology Review*, deepfakes and the technology that creates them, generative adversarial networks, were created over a beer.[2] In 2014, doctoral student Ian Goodfellow went drinking with several of his classmates. They were celebrating the graduation of a classmate. Over pints that night, Ian asked his friends for help. He needed a way for a computer program to generate photos by itself. The applications of computer-generated synthetic data were massive. These images could be used to fill holes in data collection with synthetic data to help remove collection or selection biases. Synthetic images could also be used to drastically lower collection costs for research projects, allowing the cash-strapped doctoral students to do more with their paltry budgets.

Ian was not the first to think of using machines to generate synthetic data. Academic and industry researchers had been using

AI to generate images and other data types for several years. The problem was that these were not very good. Images would come out blurry. Faces would be unrecognizable. To solve this, Ian's doctoral student friends suggested a statistical mapping of elements of the images to allow the machine to create them on its own. However, the challenge with this approach was simply that it would take up too much time and the computing costs would be too high.

Ian decided he would try a different approach to save time. He would point two neural networks against each other. One AI would create an image. The other AI would try to detect if it was real or not. The two AIs would be adversaries, with one creating fake images and the other detecting them. The AI creating the image would in turn be rewarded for getting more and more realistic, using an advanced AI technique known as reinforcement learning. If the AI creating the image was good enough to fool the AI scanning for fakes, it was likely to fool a human also. Ian liked this idea partially because it made some sense. In recent years, neural networks had become extremely good at detecting real versus fake images. This could allow him to generate insight at machine speeds. Ian also liked the idea because he was in the midst of drinking beers and wanted some results by morning. This would allow him to code for a few hours while his girlfriend was asleep and then let the machines do most of the work.

When he awoke, so the legend goes, the first deepfakes were available. He had created an AI that was able to generate images that could beat another AI's real-or-not-real detection capabilities. Ian's methods became known as generative adversarial networks, shortened to GANs. They were named so because they generated new data based on an adversarial relationship between the creating AI trying to fake the detecting AI. This seemingly simple technique turned into a significant risk within information and disinformation warfare and propelled Ian to becoming an AI celebrity. The most important reason why GANs are important is because they transformed AIs into something that could only detect into something that can also create. In essence, Ian gave AI systems creativity. When speaking to *MIT Technology Review*, Yann LeCun, Facebook's chief AI scientist, has called GANs "the coolest idea in deep learning in the last 20 years." Another AI luminary, Andrew Ng, the former chief scientist of China's Baidu, says GANs represent "a significant and fundamental advance" that's inspired a growing global community of researchers.

A Deepfake is the outcome of a back-and-forth rivalry between two AI systems. The most common analogy used is between an art forger and an art expert who is detecting forgeries. Let's imagine the two have a friendly competition to see who can outsmart the other. The forger starts with a bad forgery that is easily detected. He eventually gets so good at forging that he tricks the art expert into thinking one of his creations is the real thing. The forging AI in a GAN is called the generator. The art expert is called the discriminator. Both of the AIs are trained on the same initial training dataset. The rivalry continues until the generator is able to outsmart the discriminator. The output is a deepfake.

GANs, and the deepfakes they create, have opened up entirely new worlds for AI. Now, AI can compose music, digital art, and even poetry. GANs are less reliant on human programmers to tell the machine exactly what to do and what is in the training dataset, opening up significant opportunities in the field of unsupervised machine learning. This in turn opens up huge opportunities in commercial solutions for self-driving cars and other autonomous vehicles, where labeled training data required for supervised machine learning has been increasingly challenging to obtain given the volume of data and multitude of objects on a road to tag.

At the same time, GANs open up new areas for disinformation campaigns. As useful as GANs are to AI research and the advancement of AI capabilities, they bring with them significant societal risks. Part of these risks fall under the category of existential risks. GANs are able to reason with closer to human-like consciousness. This also means that the artifacts of GANs can fool humans. This is where they become dangerous. Deepfakes have all the hallmarks of a real image, video, or audio segment. But they are not real. Technology that can create artificial events that can be widely spread represents a significant threat to targeted organizations.

CHAPTER ELEVEN

Will Someone Hack My AI?

SAME THREATS, NEW TECH

WILL SOMEONE HACK MY AI? THE ANSWER IS, IT DEPENDS ON who you are, how much is at stake, and how hard it is. So, in short, maybe? The threat level at which your model will be under attack depends on a lot of factors, including the impact of the hack, the difficulty of the hack, and availability of your model. It should go without saying that the damage done by a hack on an AI marketing bot will be significantly less than a similarly executed attack on a military AI weapons system. While in theory all models can be hacked, it is unlikely that many of them will because both the risk–reward and difficulty–reward ratios are too high.

Leaders and AI development teams need to understand and score a model's risk profile in order to prioritize both model hardening and security measures and accurately estimate AI security budgets. To complete this, I propose an AI threat model as a framework to work from. This framework is not meant to be exhaustive, and organizations can develop their own frameworks to work from. But it will give organizations and AI development teams a starting point as they begin to think about AI riskiness.

SECURE AI LIFE CYCLE

Threat modeling methods are commonplace in the cybersecurity industry. Although there are many types, with various industry bodies and security professionals choosing the one they feel most appropriate for the job, threat models all tend to have the same underlying components. First, threat models typically involve looking at a cybersystem in abstract, without thinking about the specifics of a system. For a cybersecurity analyst, the system might be as general as underlying customer data or the robotic controls at a manufacturing plant. Second, the motivations of the attacker are taken into account. This is sometimes referred to as the pay-off, or expected outcome, from the attack. Finally, all risk models include a focus on the methods used to carry out the attack. These three elements can be thought of simply as Why, Who, and How.

The best use of the Why, Who, and How threat modeling is done early in the development life cycle. Thinking about security requirements should not be an afterthought but instead should be carefully considered at the onset of development and tested throughout. This can lead to proactive decisions regarding risks and trade-offs to architectural decisions that allow for threats to be reduced from the start. In recent years, the active integration of security testing into software development has taken on the name secure development life cycle, or SDLC. There are many SDLC frameworks, with companies including Microsoft[1] and government agencies including the National Institute of Standards and Technology (NIST)[2] each having their own version. The implementation of SDLCs has helped standardize the process of secure software development and greatly improved the security of many organizations. SDLCs, when used correctly, can be thought of as the action or process associated with threat modeling.

Today, data science and AI development teams lack a secure AI life cycle (SAILC). Security in the data science community is where cybersecurity was twenty years ago. At best, it is an afterthought, and most of the time it is barely considered at all.

Granted, AI use today is still limited. There is a popular joke in the data science community that AI is for PowerPoint presentations while machine learning is in python (and is sometimes even used!). The boundaries on practical AI implementations limit both the opportunity for an attacker to attack an AI as well as the potential payout. Meanwhile, the technical know-how to hack AI systems, mainly the effective use of operationalized adversarial machine learning, is still a relatively new field. This leaves few would-be attackers with the skills to even carry about an attack, if one were possible.

But these caveats should not dissuade organizations from investing in and implementing SAILC capabilities. From smart voice assistants to self-driving cars to a range of business analytics platforms, AI is increasingly found at the heart of strategic initiatives. Leaders in all industries are actively experimenting with AI and more than half of CEOs believing it is core to their strategic position in the future.

For data scientists and AI developers, security is generally an afterthought if it is brought up at all. If you ask an AI developer to describe their job, she may talk about generating insights, building new products and services, creating optimization workflows, or building next-generation platforms. If she includes security in their job descriptions at all, it will almost always be somewhere near the end of the list and usually in the context of traditional cybersecurity protocols.

This is not the fault of the AI developer. Beyond the fact that cybersecurity budgets are already stretched thin, for many people, thinking like an adversary can be unusual. For instance, the engineers who designed the early generations of self-driving cars did not expect their cars' computer vision systems to be hacked in ways that can cause crashes. Indeed, the very purpose of these systems was to avoid crashes! But this is exactly what happened. Likewise, the creators of voice applications did not write their code thinking people would embed whisper data into audio files in a way that would compromise the applications' security, but again, this is exactly what came to pass. Adversaries and AI hackers take advantage of the optimism of system developers and use their lack of security to their advantage.

Such optimism-to-a-fault has shaped the tech space since at least the advent of the Internet. Few people anticipated that the Internet would become the driving engine of global finance and communications (and so much more) it has, and as a result, few meaningful security measures were built into its foundation. This created a massive cybersecurity debt that organizations are still trying to pay off today. The harsh reality is, we're building AI in much the same way—it is open and rarely has security built in.

Speaking on business interest in AI today, IDC Research Director for Cognitive/Artificial Intelligence Systems David Schubmehl says, "Interest and awareness of AI is at a fever pitch. Every industry and every organization should be evaluating AI to see how it will affect their business processes and go-to-market efficiencies."[3]

This interest in AI is leading to increased investment in AI, which is leading to increased deployment and use in critical mission and business settings. The common critique that not enough AI hacking is done as of today should not negate the development and implementation of a SAILC. If your organization is considering deploying AI in any context relevant to your strategic vision, a SAILC is needed to prevent AI development from facing the same challenges that plagued the cybersecurity industry. As of the summer of 2020, the average cost of a traditional cyberbreach is roughly $116 million per breach in the United States[4] and $3.92 million globally.[5] This translates to billions of dollars spent every year by organizations after the breach has already occurred. Many of these losses could have been avoided had proper SDLC protocols been followed.

AI security and risk mitigation is still in its infancy. But instead of waiting for threats to emerge, as industry practitioners ultimately have done with most digital technologies, it is better that we take pragmatic steps today. The security vulnerabilities and attack vectors of AI are still being explored. But we have enough information today to make a concerted effort toward security by design. This will not only save companies millions in the future by making it harder for would-be attackers to succeed, but given AI's applications in healthcare, transformation, and military applications, such actions will also save lives.

A threat model underpinning the process of a SAILC has many of the same underlying components of traditional cybersecurity threat models. The AI threat model that I have developed includes both a threat space (the understanding of the anatomy of a potential attack) and the mitigation space (an analysis of best techniques to limit the threat). I purposefully avoid using the team's problem space and solution space in this framework, despite these terms being common in threat modeling language. This is because thinking of any threat as a solved problem can yield improper thinking. Threat landscapes, especially on the AI threat surface, are constantly changing. This is why implementing a SAILC is so critical. If done correctly, this process will be an iterative cycle, as opposed to a stand-alone solution.

Why

The first question that must be asked is in a SAILC threat model is why will someone attack the model in the first place? This question is really about the motivations of an attacker. While there may be academic or bragging value in hacking a system just for the sake of doing so, in reality most hacks have a financial, competitive, or national security motivation behind them. Understanding why someone would want to disrupt your AI is important. For example, even if both systems use sophisticated computer vision, it is more likely that an adversary will want to disrupt the AI of a weaponized military drone in order to hide their forces, as opposed to hacking a restocking robot in a grocery store. The same

SECURE AI LIFECYCLE THREAT MODEL

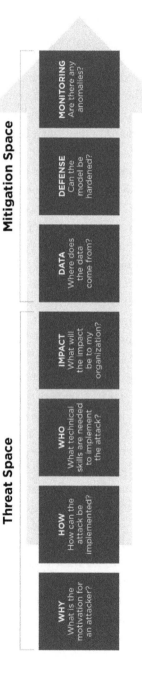

Threat Space

Mitigation Space

WHY What is the motivation for an attacker?

HOW How can the attack be implemented?

WHO What technical skills are needed to implement the attack?

IMPACT What will the impact be to my organization?

DATA Where does the data come from?

DEFENSE Can the model be hardened?

MONITORING Are there any anomalies?

Figure 11.1. Diagram of the parts of the secure AI life cycle.

is true within the same organization as well. For example, stealing or gaining access to underlying consumer financial transaction data at a big bank is going to have a bigger payout than hacking that same bank's marketing bot on Twitter.

An important consideration behind why an attacker might seek to hack an AI is not even a successful attack. Instead, an adversary might simply be trying to sow mistrust in the AI system and prevent its use. This is especially true in national security contexts where eroding trust in an AI might be the primary reason to attack an AI system in the first place. Tyler Sweatt, a U.S. Army veteran and technology expert, is deeply concerned about this new field of psychological warfare. I was able to meet up with Tyler near his home in northern Virginia. He looks the part of a U.S. Army veteran, barrel-chested and bearded. He has a booming voice, which makes you think he is yelling even when discussing highly technical topics. "I started my career as a bomb guy. Later did intelligence work," he tells me. Since leaving the army, Tyler has been building cutting-edge technology applications for the military and national security industry. He has been the primary voice behind this new field of anti-AI psychological warfare. He is worried that even just the threat of a successful attack will force the U.S. military to kick AI systems offline. This could render AI capabilities offline not due to a technical issue, but because of mistrust.

He says, "AI will permeate all aspects of daily life in coming years, from security to health to social and more. Having the ability to sow distrust between a society and the AI powering it will be a critical tool kit in government around the world. The inability to defend against such attacks will kick some governments out of the digital age." Tyler sees a new psyops frontier between those who want to use AI and those who want to sow distrust in those systems to prevent their adversaries from using them. For example, the United States might release information that it has weaponized tactical AI attacks in new battlefield cyberunits. This could cause the Chinese military to switch off their AI at a critical time, increasing the U.S. military's ability to disrupt its kill chain in a hypothetical great power conflict. The same is of course true in the reverse. The threat that an AI is susceptible to a successful attack can prevent their use. This is important to remember when viewing whether an attacker will try to hack an AI. AI attacks do not need to be successful to be effective.

How

After knowing why, you have to understand how. Will this be an evasion attack? Data poisoning? Model inversion? How will an attacker gain access to my model? Will there be additional cybersecurity hacks involved? The how is really where the imagination of the SAILC team comes into play. Will there need to be a physical site breach to gain access? Can the AI endpoint be attacked using off-the-shelf tools? It is important here to think through all plausible options, even those that seem out of the ordinary. If the prize is big enough, an AI hacker will be willing to try nearly anything to get to the prize.

Understanding the how is more complicated than it may seem, primarily because it requires a change in thinking by the AI development or data science team. The best defense is a good make-believe. In order to know how an adversary might want to hack your model, it is best to think like an attacker. Developed in the military, the concept of red teaming is the process of forcing a team to think through dissenting strategic or security elements. In other words, red teaming involves assuming the role not only of the devil's advocate, but of an attacker. Red teaming a model is the process of understanding exactly how an adversary can pull off an attack.

High-performing companies like Amazon and Google frequently use red teaming (sometimes under a different name) to assess new strategies, products, and services. According to legendary Ford executive Alan Mulally, red teaming is essential at a strategic level "because your competitors are changing, the technology is changing, and you're never done. You always need to be working on a better plan to serve your customers and grow your business."[6]

In more strictly technological contexts, red teams adopt the optimism of a technologist while assuming the role of an attacker. A good red team should neither shoot down a project nor stand in the way of a new product rollout, but should instead take for granted that attackers will find some way to trick, fool, or maliciously abuse a digital system—and try to facilitate a defense against such abuses. Simply put, red teams look at new technologies and ask, "If I were an attacker, how might I use this technology to my advantage?"

Apart from those who find weaknesses for a living, such as military members or security professionals, most people don't go about their everyday lives thinking about how something can be attacked or exploited. In the digital world, naivete regarding the likelihood of an attack can lead to vulnerable systems. Too often, in software development security checks and patches are only completed at the end of a development cycle by a separate team. While recent trends toward DevSecOps and similar frameworks are helpful, too often security remains an afterthought.

Given the increasing use of AI in critical mission and business systems, security cannot be an afterthought. The practice of continually red teaming your model to both understand an adversary as well as determine the likelihood of a successful attack must be completed as a continual part of the secure AI development life cycle.

In every case I have examined, including banking, insurance, self-driving vehicles, and military AI systems, successful attacks can be found. But the goal of this part of the process is not only to find successful attacks but also to gain an understanding of how an adversary was able to create the attack. For example, were there other systems the adversary needed access to? What were the computing costs and perturbation distance required, and is the attack payout high enough to convince an adversary to spend on creating the attack?

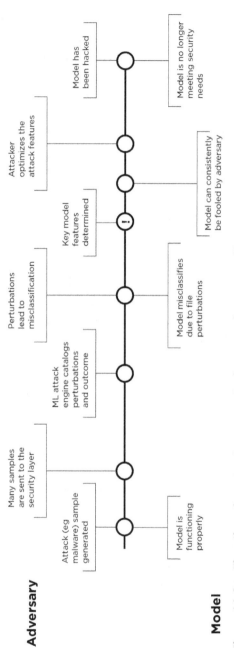

Figure 11.2.   Time line of an attack on an endpoint protection system AI.

Who

After they have identified, an organization needs to understand who would do this. This is not so much an exercise in identifying possible criminal groups or state-level adversaries. For example, the U.S. military and intelligence community already know that China and Russia will try to hack their AI. The question is more importantly about the skill levels involved. Is a PhD in statistics or machine learning needed to piece together the intelligence gathered? Or can this hack be carried out using open-source tools pieced together by someone only moderately familiar with AI?

During the who stage, open-source intelligence on adversarial machine learning is critical. There is a rapidly emerging body of online resources available for those who want to hack AI systems. Some are found on the dark web and are for hacking specific things, like a leading auto manufacturer's self-driving vehicles. Others are provided by big companies, like IBM, as a research tool.[7] In hacking circles for cybersecurity, practitioners sometimes pejoratively describe persons who use off-the-shelf tools as "script kitties." But this pejorative language doesn't make those who use the rapidly increasing pool of open-source AI attack libraries any less dangerous, especially when there are no defenses in place. Continually updating open-source intelligence on the current trends and available libraries in adversarial machine learning will help organizations understand the personalities needed behind the threats they face.

Impact

Finally, an organization needs to understand the impact of the hack. Some impacts are straightforward. If an enemy is able to trick an intelligence, surveillance, and reconnaissance drone's AI system that its troops are friendly troops or not there at all, the impact can be fatal to service members and decisive on the battlefield. Meanwhile, AI systems that leak classified, sensitive, or regulated information could lead to significant fines for the company if they are in the healthcare, financial services, insurance, or other heavily regulated sector. Some impacts might also be quite minor, such as a marketing blunder or better online shopping deals for a user to optimize his behavior to match a marketing bot. The impact of an AI hack is going to differ from industry to industry and also from use case to use case. It is best at this stage in a SAILC to involve not only the data science team, but also legal, risk, compliance, and business continuity specialists when reviewing the final output.

Those first four elements—why, how, who, and impact—make up the threat space of the AI threat model. Each of these need to be considered in a risk-adjusted manner to make up the final prioritization of the biggest potential AI security risks to an organization. Once the biggest threats have been identified and prioritized, the team needs to be able to mitigate them to the extent possible. This is where the second part of the SAILC threat model comes in, the mitigation space.

Data

First, it is important in AI to start with the data. As we have seen, data providence and dataset solutions are critical to building robust AI systems. A thorough analysis should take place when the threat model indicates that data poisoning or data tampering, including evasion attacks, can take place. It is possible to mitigate data threats through careful monitoring of training and data ingesting.

Defenses

Next, the team will want to consider model defenses. There is often a trade-off between model precision and accuracy and the robustness of a model to adversarial attack. It is in this stage that these trade-offs need to be carefully considered. For example, some models such as those used in precision medicine require high fidelity and are unlikely to come under attack. Others, such as models used in a military context, may require an adversary-robust model, which will have lower precision scores during training. Models that are in low-threat environments do not need

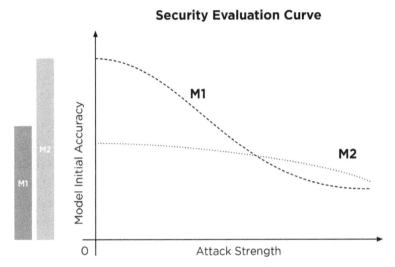

Figure 11.3. The trade-off between classification accuracy and security in AI.

to face these trade-offs, but they are critical for the successful implementation of models in high-threat environments.

Monitoring

Finally, AI systems cannot simply run without ongoing monitoring. This is true for all AI systems, as some ability to monitor their computing costs and data access is necessary for simple cybersecurity and IT protocol. For high-threat AI systems, active monitoring and local interpretability tools are likely required to ensure that any anomalies in usage or in feature importance are flagged to a human operator. These could be indicators of an AI attack and should be investigated. These monitoring capabilities can be fed into a security operations center (SOC) as opposed to keeping them with the data science team. This will require a culture shift where the data science and AI development teams at large organizations are not the primary professionals responsible for monitoring AI.

The end result of a SAILC process is a prioritization of AI risks at an organization. These risks and mitigation strategies need to be effectively communicated both to the AI development and data science teams and also to the chief information security officer and to the model risk management team. Therefore, part of the job of AI development teams is translating between the various cross-functional parts of their organization and effectively communicating not only what risks are present, but how to fix the ones that can be fixed and mitigate the effects of AI risks that cannot be. A properly implemented SAILC team should not become a barrier to the acceleration of AI applications but instead be a necessary component in the acceleration of safer AI deployments.

Chapter Twelve

The Machine Told Us to Do It

Our Current Tools Are Not Enough

It was a hot day in the American South. It was humid enough that the walk from the parking lot to the front desk of our client left me dripping with sweat. I paused in the lobby right under an AC vent, checking unimportant emails on my phone while making them look very important as a way to stall. I was extremely happy that my colleagues and I were not expected to be in suits or button-downs. We were the AI team, after all, and our black T-shirts and flip-flops fit the bill. We had been brought down to the sticky South because things were moving slowly. We joked in the car ride over that it was because of the heat. But in reality, it was because the AI was not explaining itself.

In the conference room my client, a large financial institution, went over the challenge. As a financial company with millions of consumer accounts, they were heavily regulated. In the past decade, the firm had invested heavily in technology as a way to cut back on costs, increase security, and provide new and faster services to their customers. Many of these automations were routine tasks, meaning there was no advanced statistics, machine learning, or AI involved. They were simple if-then statements and corresponding logic trees. But while these automated systems were not complex, they were very complicated. Sometimes the automated task involved multiple teams in several departments. Others involved setting risk on traders across the firm by looking at underlying market conditions and valuing their balance sheet "at risk." And automated systems helped detect and prevent fraudulent transactions. To ensure compliance across these rapidly expanding automated systems, the firm had invested in a robust model risk management team. This team was made up primarily of legal experts and financial analysts and was located within their compliance department. As those in finance will be quick to tell you, the compliance office is not usually the most innovative part of the firm.

The model risk management team had grown in importance over the last decade. In 2011, the board of governors of the Federal Reserve had issued Supervisory Letter SR11-7, which had elevated the importance of model risk in the financial systems. The supervisory letter states that "banking organizations should be attentive to the possible adverse consequences (including financial loss) of decisions based on models that are incorrect or misused, and should address those consequences through active model risk management."[1] This was designed to build on earlier references to the model risk management dating back as far as 2000. That year, the Federal Reserve had focused primarily on the need for increased model validation. Supervisory Letter SR09-01 highlighted "various concepts pertinent to model risk management, including standards for validation and review, model validation documentation, and back-testing."[2] In addition, the Federal Reserve's important Trading and Capital-Markets Activities Manual also discusses validation and model risk management.[3] And while the Federal Reserve had led the charge for model risk management, other regulatory bodies, including the Federal Deposit Insurance Corporation (FDIC), had their own guidance for banks. For example, in FIL 17022, the FDIC provides regulatory guidance for "model development, implementation, and use; model validation; and governance, policies, and controls."[4] These specific risk areas are addressed in detail throughout the document and are designed to assess if an "institution's model use is significant, complex, or poses elevated risk to the institution."[5] Other financial regulatory bodies, such as the Office of the Comptroller of the Currency (OCC), also have released guidance on model risk across areas, including "underwriting credit; valuing exposures, instruments, and positions; measuring risk; managing and safeguarding client assets; and determining capital and reserve adequacy."[6]

These overlapping, and at times confusing, regulations are well intentioned. In recent years, the financial services industry has been one of the fastest adopters of automation, ranging from the basic automation of routine back-office tasks to sophisticated financial instrument trading done at leading hedge funds. The OCC in particular took note of the far-reaching impact automation was having on the financial services industry, noting that while "models can improve business decisions, they also impose costs, including the potential for adverse consequences from decisions based on models that are either incorrect or misused. The potential for poor business and strategic decisions, financial losses, or damage to a bank's reputation when models play a material role is the essence of 'model risk.'"[7] Model risk management, in the broadest view, was designed primarily to avoid outsized risks based on model behavior.

In the aftermath of the 2008 financial crisis, financial institutions became incredibly concerned about the risks associated with models, with good reason. Many economists, regulators, and market pundits blamed the banking industry's overreliance on models. These models, especially those predicting the value of financial derivatives based on housing prices and associated mortgage payments, were based on faulty assumptions of risk associated with selection bias on historical trends. Even though the models were statistically sound based on their data, ultimately the models in use leading up to the 2008 crisis failed to account for a nationwide collapse in housing prices and associated spillover effects.

Entire books, dissertations, and MBA courses are now taught on the banking industry's failure to adequately manage risk using models, so I will avoid going into too much detail here. What is crucial for our purposes is that these models were primarily statistics-based models. Very few, if any, were machine learning or AI. The aftermath of the 2008 crisis was a wake-up call to regulators and banks wanting to avoid costly fines for noncompliance to implement these new regulations with earnest. Some financial institutions built entire teams of model risk management professionals while others built model risk management working groups made up of cross-functional professionals from legal, risk, compliance, and market teams.

All of these model risk management teams shared a common trait: they were focused primarily on statistical models and purely deterministic systems. A deterministic system involves no randomness in the development of future states of the system. Therefore for any possible input, there is a known output. Model risk management teams are primarily focused on evaluating all possible inputs and their associated outputs and often required as close to an if-then scenario as possible. And in purely statistical models and robotic process automation, model risk management based on regulatory guidance has worked quite well. But then came AI.

"So after my team is done creating the model, testing it, and running a few live scenarios with other datasets, we send it to compliance. They have a model risk management protocol to go through," my client said. She is an experienced cybersecurity professional who had recently taken ownership of the firm's internal data science projects for fraud detection. Beyond being an accomplished cybersecurity lead, she also has advanced degrees in data science and has led AI development teams in the past. She is one of those rare professionals who seamlessly blended two critical fields at exactly the right time.

"After it goes to model risk management, it sits there. Usually for a few weeks at a time. Then we get questions back. These questions are usually about specific possible data inputs. So, we run it through the AI model. And report back. Then we get questions a few weeks later asking why we got those answers," she explains. She is walking me through their workflow. They had dozens of professionals working as data scientists and AI developers across the organization. She oversaw not only data scientists working on fraud, but also on other areas of automation and insight generation. As a talented, experienced professional running a sophisticated and highly paid team, I was expecting her organization to be deploying AI rapidly. She clearly had the support of the firm's leadership. In their annual earnings report the company's CEO had mentioned AI and automation several times during his call with the investment community.

"At the end of the day, every time we send a model to model risk management for approval it turns into two things. First is a random walk down specific inputs that they want to see outputs for. Fine. This is time consuming but doable. Then send a list of inputs, we put it through the AI engine, and send the results back. Second, we need to basically teach them a Master Class in AI theory to get them comfortable with each new model. It never really works." What she is describing is the process it takes to get an AI model through the model risk management framework at her firm. On average, it takes four to six months to have AI models approved.

"The compliance team was always concerned that they would have to tell a regulator, 'well the machine told us to do it' and we would be fined," she told me. "The reason they wanted to see all of the possible input and output combinations possible was because they felt that was the only way they could tell a regulator that the machine was compliant. They were scared of the black box AI and were even more scared that the regulators were scared of the black box."

As we dug deeper, my colleagues and I discovered that it was not due to poor documentation or justification for model architecture on behalf of the data science team. Unlike many companies I have spoken with, this client in particular already had a robust model assessment criteria and information sharing in place. The challenge was that it was created by data scientists and AI experts for other experts. These experts understood the trade-offs between, for example, random forests and deep neural nets. And they carefully constructed their models to be the right model for the job. The breakdown in communication came when they needed to justify these findings to the model risk management team.

A similar situation unfolded when talking with a leading insurance company. This firm, known for being on the bleeding edge of insurance technology, was actively using AI to make underwriting decisions on home and business property. On the surface, it looked like they had an obvious edge and I could not figure out why their competition was not doing the same. The challenge, however, was nearly identical.

"It was one of the big states. Think New York, Texas, or California. And their insurance regulator, the Department of Insurance, wanted to certify that our model was compliant," said the insurance company's director of growth. He was walking me through the challenges they faced when using AI models for underwriting. "Ultimately, they asked us to provide outputs for something like fifty thousand different addresses. And then a random sample from every zip code in the state. We complied and ended up shipping them something like twenty thousand printed pages of documentation."

Compared to many regulatory bodies, the state's Department of Insurance was seemingly well equipped to examine an AI model. After all, insurance companies and their regulators are well versed in the statistical models that underpin insurance practices. However, even these mathematically sophisticated organizations are unable to adequately assess whether a model is of high quality and compliant. The only methods at their disposal were to throw lots of data points at the AI and individually screen what the output is.

When speaking about the challenges in the insurance industry in particular, Amir Cohen, the cofounder and CTO of the AI development firm Planck, boils everything down to trust. "AI will suffer," he says, "at the beginning from lack of trust. Change is never an easy thing for people or organizations, and as it will permanently change processes that [insurance companies] have been running manually for decades, the switch will not happen in a day."[8] The trust Cohen is referring to is a trust that a model will perform and is secure. What Cohen touches on only briefly is the underlying cultural shift from manual operations to reliance on automated, nondeterministic AI systems. What he calls trust is really the difference between manually checking for all possible inputs and outputs versus understanding and evaluating the logic of an AI, where for any input the output is unknowable at the onset.

The challenges my insurance and financial services clients face when it comes to assessing AI model risk are nearly identical. Mainly, AI development teams are unable to effectively communicate an AI's legality and security to compliance-based organizations. Compliance organizations in this case refer to both regulatory bodies, such as a state-level department of insurance, and internal legal, risk, and compliance teams. This inability to effectively communicate between teams causes significant time delays on model implementation while wasting both time and resources. This is not the fault of these compliance organizations. In the previous chapters we have discussed in depth several ways that models can fail based on bad inputs and adversarial action. Regulators and risk teams alike are realizing that AI systems can be hacked. Rules-based automated systems, which the financial services model risk management and the insurance regulators are used to seeing, cannot be hacked using these methods. For any input, there is a known, mapped output. It is AI's nondeterministic nature that makes it both vulnerable and opaque. AI can also fail because it is poorly constructed, ill-suited for a specific use case, and AI performance can change over time. With all of these risks, regulators and compliance teams have good reason to be concerned about the performance, bias, and security of models.

Although both insurance and financial services industries have a direct impact on the lives of their customers, it is unlikely that an AI's hacking or failure in either industry will result in lives lost. This is not the case in other sectors, such as aviation and defense, where the hacking or failure of an AI system can have life-and-death consequences. In light of these risks, it would be logical to assume that these industries have already established a robust tool kit to assess the security and performance of their AI systems. But this is not the case.

In late 2019, the RAND Corporation was commissioned by the U.S. Department of Defense's Joint Artificial Intelligence Center (JAIC) to write an assessment of the JAIC's efforts in AI. Established in 2018 and administered under the DoD's chief information officer, the mission of the center is to establish a common set of "AI standards tools, shared data, reusable technology, processes, and expertise" for the entirety of the DoD.[9] Underpinning the creation of the center was a single fear: the United States could potentially lose its technology edge in AI to Russia and China. The JAIC's mission goes beyond just establishing a set of tools and technology. The mission behind the mission, so to speak, is to maintain the U.S. military's information technology superiority in great power competition.

The first major, publicly announced project undertaken by the JAIC was the controversial Project Maven. Officially called the Algorithmic Warfare Cross-Functional Team, Project Maven had been launched the year prior in April 2017. According to a Pentagon spokesperson, the project's mission is to create "computer-vision algorithms needed to help military and civilian analysts encumbered by the sheer volume of full-motion video data that DoD collects every day in support of counterinsurgency and counterterrorism operations." Project Maven was intended as, and remains, highly sensitive, and most information about the project remains classified. What is known about the project publicly is that it is an advanced program to use computer vision AI to sort through the massive amounts of livestream videos and other information captured by the massive U.S. military intelligence, surveillance, and reconnaissance apparatus. This means that AI systems are being used to identify persons, including enemy combatants, and track their movements. Among the first usable AI to be shipped to a warzone was a computer vision system that identifies "38 classes of objects that represent the kinds of things the department needs to detect, especially in the fight against the Islamic State of Iraq and Syria,"[10] according to the DoD.

Reasonably, this means that AI is actively in use by U.S. warfighters in active combat zones. What members of the defense community and even members of Congress believed was that due to the potential lethality of these systems as part of the intelligence collection and exploitation cycle, JAIC and Project Maven would have significant security and risk management protocols in place. What the RAND Corporation found, however, was the opposite.

"The field is evolving quickly, with the algorithms that drive the current push in AI optimized for commercial, rather than Defense Department use. However, the current state of AI verification, validation and testing is nowhere close to ensuring the performance and safety of AI applications, particularly where safety-critical systems are concerned, researchers found."[11]

Despite the potential for lethal action associated with AI deployments on the battlefield, the use of the technology should, in theory, yield significant quality and security assurance testing from the military. At present, the benefits should outweigh the risks. But RAND found the opposite. Instead, the U.S. military is running into the same challenges as the private sector. The study found that AI security, in particular for U.S. military applications, lacks the same rigorous cybersecurity protocols that traditional software must use when it is being deployed. This is the same problem facing private sector companies as well. Despite significant experience in assessing and mitigating risk from complex software, logistics, and mechanical systems, the U.S. military is not well positioned to assess the underlying security and performance risks of AI, even at centers dedicated purely to the acquisition of these technologies into warfighting capabilities.

The inability of organizations to rapidly assess AI performance and security does not illustrate all of the challenges of implementing AI. There are massive technical, workforce, and cultural challenges as well. On the technical side, AI strategies run into implementation challenges ranging from IT readiness, to data preparation, to traditional privacy and security concerns. These are coupled with an ongoing workforce shortage for advanced data science and AI developers and managers who can effectively lead AI-centric organizations. Finally, organizations must undergo cultural shifts as well. There are trade-offs to be made by implementing AI, not least of which includes potentially shedding jobs as automation increases. In many ways, these technical, workforce, and cultural challenges are the biggest barriers to successful AI implementation. These barriers represent more significant hurdles to date than security.

What these examples from financial services, the insurance industry, and the U.S. military illustrate is that many of our most sophisticated organizations are ill prepared to understand, let alone mitigate, AI performance and security risks. AI security and quality assurance is considered an afterthought, if it is considered at all. This leaves these organizations open to an AI hack and also can put lives at risk. All organizations, whether they are just beginning their AI journey or already have robust AI development teams in place, must adopt a riskcentric AI development and model risk management framework to ensure the quality, security, and compliance of their AI models.

ASKING THE RIGHT QUESTIONS

In assessing organizations' AI pipelines across financial services, insurance, government, and even consumer sectors,[12] one critical piece of information stood out. They were not asking the right questions. Meanwhile, different teams, such as data science and legal teams, were not asking the same questions, which creates muddied responses and failed communication. The data science and AI development teams were primarily focused on data science metrics, such as precision, accuracy, and F1 scores. These are all critical metrics in assessing the quality of a model. However, in a vacuum they do not answer the operational and compliance questions that the cross-functional team members of model risk management were asking. These questions concerned the legality, bias, and performance in so-called tail risk events. Basically, cross-functional teams were asking for justification as to why a model was sound and wanted to know when it would fail.

A modern, AI-ready model risk management solution incorporates the secure AI life cycle (SAILC), discussed in chapter 11, and layers on critical questions an organization must answer to validate a model's quality, security, legality, and ethics. When done correctly, AI model risk management seamlessly combines cross-functional reviews, reports, and ongoing monitoring of an AI from initial data collection to successful implementation.

Although this approach is far more comprehensive than compliance reviews at most large firms, AI model risk management should not slow down the use of AI. On the contrary, successful implementation of AI model risk management should accelerate AI experimentation, testing, and adoption at enterprise scale. This is because by baking in answers to the critical quality, security, and compliance questions from the beginning, final validation of AI systems will be straightforward. If an AI model risk management process has been followed from start to finish, all questions will have already been answered in a documented, auditable trail. This will require a significant cultural transition as cross-functional legal, risk, and compliance teams will need to work more closely with data scientists and AI developers. This is likely to cause friction at first as these organizations learn to work together.

Assessing AI using model risk management processes includes three primary categories: AI quality, security, and compliance. Each of these requires different types of reporting and metrics. Today, no single set of tools is available to manage an AI model risk management workflow. Instead, organizations must rely on a collection of tools and internal processes to manage the ongoing process of AI model risk management. The following sections are not meant to be in linear order, as components of each happen across the model development life cycle.

The standards that organizations put in place across quality, security, and compliance will vary from AI use case to use case. A computer vision AI that predicts same store sales revenue estimates for Wall Street by looking at cars in a parking lot does not have the same security or legal requirements as a computer vision AI that diagnoses cancerous cells. Every question asked of an AI across its model risk management cycle must be appropriately contextualized. The same is true even within high-security environments such as the military. For example, one of the largest unclassified uses of AI by the U.S. Air Force is in predictive maintenance. The USAF uses connected sensors on its airframes to predict failure in advance, leading to decreased time spent in the shop and lowering replacement costs. While an adversary could likely find a way to disrupt these predictions, their payoff would be limited. However, the USAF also uses AI in sophisticated weapons and targeting systems. These will almost surely come under attack by an adversary and also carry a much higher cost for failure. If a predictive maintenance AI fails, a plane might be grounded. At worst, it will be grounded at a critical time. However, if an AI weapon system fails, people may die. Because of the massive difference in the impact of an AI's failure and performance, organizations should designate specific thresholds for types of use cases. These thresholds should take into account the underlying quality, security, and compliance risks.

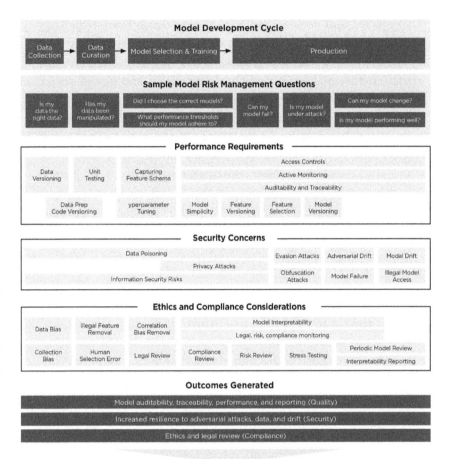

ACCELERATED ADOPTION OF AI

Figure 12.1.   An example of an AI model risk management framework.

Quality, Performance, and Traceability

Starting with the model itself, the primary question that needs to be asked is: Is this model high quality? Unpacking the answer to this question primarily includes traditional data science metrics, as well as scenario planning, model simplicity testing, model stress testing, and data quality testing. All of these factors need to be completed and documented by the data science team. Following documentation, the information must be shared internally to the data science team for expert review. Following data science review, each incremental evaluation should be shared and signed off on by members of the cross-functional model risk management team.

Model quality refers to the appropriateness of a model and its underlying data for a certain task. Model performance is the model's ability to perform not only on training data but also in real-world scenarios. These two characteristics of an AI, that it is the right model for the job and that it performs past a certain threshold, are both building blocks of AI model risk management and are typically where data science teams spend most of their time. Common performance scores, including AI accuracy, precision, and F1 scores, among others, are useful metrics to include during these segments of AI model validation.

Traceability is often overlooked, but it provides continuity across the entire AI model risk management cycle. Traceability, sometimes called auditability, refers to the ability of any team member or manager with appropriate access to view every decision made in an AI model's development and eventual deployment. This includes who made the decisions, notations on why or what trade-offs were considered, and who from the cross-functional team approved any decisions. Traceability is key because if done correctly it forces documentation and interteam collaboration at all model decisions while creating a single shared source of truth for the AI's life cycle. That means when it comes time to push the AI into a production environment, all of the information is already codified, documented, and approved.

The best examples of traceability come from inside the U.S. intelligence community. At the CIA, case officers, those responsible for recruiting and running spies in foreign countries, routinely document as much detail as possible at every step.

"We make it so that any other officer can easily pick up where we left off," a retired career intelligence officer told me over a quiet

beer in northern Virginia. We had met up to discuss digital transformation trends across the U.S. government. The conversation
had quickly turned toward what we both saw as an impediment
in AI usage within national security: poor traceability within data
science teams. "You never know what's going to happen in the
field. You could be compromised and have to leave the country.
You could be reassigned. If this is in a hostile territory, you might
even be killed or wounded," the aging officer told me. "For that
reason, we make it so no matter what happens your work can be
continued immediately by someone else with as much information as you have."

While not nearly as dramatic or dangerous, data science teams
should take the same approach. Data scientists and AI developers
are in high demand and many shift jobs after only a short period.
Meanwhile, given the numerous data science projects going on at
major enterprises, many will be assigned to new projects or teams
in the middle of working on other projects. Maintaining traceability is therefore key to maintaining continuity and avoiding
lost time, wasted effort, and a stalled project. When done correctly, traceability is not seen or used as a Big Brother approach to
quality assurance, but instead these techniques should be used to
successfully document and communicate the quality and performance of models to accelerate their use.

SECURITY AND DRIFT

The best way to avoid an AI security incident is not to be a soft
target. Therefore, a large component of AI model risk management is the implementing of a security AI life cycle (SAILC),
discussed in chapter 11. By implementing threat modeling
and security testing as early as possible in the AI development
process, organizations can avoid the trap of AI-security-as-an-
afterthought that is pervasive in most organizations. Organizations implementing SAILCs should be careful not to confuse this

process as all-encompassing of AI model risk management. This process only makes up a subset of overall risk mitigation.

At the beginning of February 2019, Gartner published a damning report on the status of AI security and risks.[13] Their report stated that "application leaders must anticipate and prepare to mitigate potential risks of data corruption, model theft, and adversarial samples." This directly calls out many of the AI hacking methodologies addressed in prior chapters. Yet, the report found that organizations were extraordinarily underprepared. The chief information security officer (CISO) at a leading bank told me, "It's not that we [the bank] don't want to secure our systems. It's that we don't know how. How many data scientists do you know who can actually build adversarial samples or understand the science? Now, how many of those people want to work at a big, boring bank?"

When Ram Shankar, Siva Kumar, and Frank Nagle conducted research for *Harvard Business Review*, they found the same pattern across organizations, including Fortune 500, small and medium-size business, and government bodies—they found that 89 percent lacked even a plan to tackle adversarial attacks against their AI systems.[14] What was most interesting to me in this study is that it was not due to a lack of awareness that there was a problem emerging. Instead, it was a lack of capability and know-how as to how to address the concern.

The authors point to two additional areas beyond internal capabilities that make it hard for organizations to strengthen their AI security posture. First, AI defenses continue to rapidly emerge. Take for example when AI security researchers found that thirteen of the leading defenses against adversarial examples from academic literature were found to be operationally useless.[15] Many organizations view security as a "check-the-box" drill. But that cannot be done with AI hacking as the field is rapidly emerging.

Second, existing copyright, product liability, and U.S. "anti-hacking" statutes may not address all AI failure modes.[16]

Some aspects of computer crime, copyright, and tort law cover some elements of perturbation, poisoning, model stealing, and model inversion attacks, while others are not covered. For example, because attacks on computer vision systems that take place in the real world, such as adversarial glasses or stickers on stop signs, do not actually give access to the underlying computer system, such attacks are not covered under the law. This means that the normal levers an organization can pull to assess and mitigate its risk were not available. Beyond legal mitigation, companies typically look to mitigate security risks through insurance. But in the case of AI, even traditional cyber insurance policies do not cover the new ways adversaries can manipulate models.[17]

Beyond successfully implementing a SAILC process, organizations that care about AI security must also do two things to fill the gaps mentioned above. First, they must elevate the role of the CISO to be a cross-functional executive overseeing the SAILC and AI model risk management. Second, they should invest in AI insurance.

A common question when discussing SAILCs with executives is: Who has responsibility for this? Which translates from executive speak to: Whose ass is on the line when this fails? A common reaction is that the design and implementation of the SAILC should be within the data science or AI development team itself. I believe that this is the wrong choice. The team and leadership most suited to tackle AI security is the team already taking on cybersecurity—the CISO. SAILCs should be managed by a cross-functional team organized by and reporting to the CISO and their staff.

This will take an evolutionary shift in the role of the CISO. Constrained by tight budgets and needing to comply with overlapping rules, requirements, regulations, and vendors, many CISOs today are overworked already. They rely on a strict set of compliance and standards. In short, except in extremely forward-leaning cases, CISOs are not thought to be the innovation engine of an organization. But in many organizations, CISOs are already taking on this role. "I got called to give a presentation to the board one Friday night," a CISO of a leading consumer brand told me. "It wasn't because of anything I did. Or even my team did. Marketing had messed up, and their marketing tool pushed advertising incorrectly. It was AI, so fell under data science. But they said it was a digital system failure, which is true, I guess. The job fell to me. Now I am in charge of making sure all of our AI doesn't crap out."

CISOs will need new tools, increased budgets, and authorities to effectively manage a SAILC. Data science teams will likely either continue reporting to business units directly or operate as a firmwide shared service. So the CISO will need to find new ways to interact with these teams and will require top cover from senior management to implement them. There is no other organization at big firms well suited to manage the combination of cyber, physical, and algorithmic risks associated with AI security. CISOs will need to hire data-fluent professionals and learn quickly about the new risks of AI and AI hacking. But cybersecurity has always been fast moving and evolving. I am sure CISOs are up to the task.

Second, organizations should also invest in AI insurance. Even when an organization does everything right within its AI model risk management framework, bad things can still happen. I developed this framework in part due to public AI failures at large organizations including Google,[18] Tesla,[19] and Uber.[20] In each of these examples, a complete AI model risk management framework does not appear to have been followed. However, these organizations and their AI development teams were still doing the best they could with the tools and processes they had. As organizations forge ahead with new AI technology implementations, it is likely that AI risks and model risk management will continue to be a sideshow, viewed as a cost center as opposed to the cost saving and AI acceleration engine it can be. This is to say that AI failures will continue to happen. As technology adoption expands, these failures will likely increase in frequency as well.

Today, cyber insurance policies will cover only a narrow band of AI risks. Namely, they will cover model stealing and data leakage. Model stealing is covered because it can be argued that there was a breach in private information. Meanwhile, data leakage is already covered and typically does not include specific language as to which attack vector was used in order to trigger the policy. But if an attacker causes bodily harm, does brand damage to the organization, destroys digital or physical property, or causes other adverse effects, cyber insurance will not cover the company.

As organizations regard AIs as more than just tools to being centers of growth and operations, having SAILCs will not cover all the risks. Mistakes and failures will happen. The insurance industry is set up to help offset these risks. It is likely that organizations looking to purchase AI insurance will need to bake in either ISO[21] or NIST[22] trusted AI standards as part of their SAILCs reporting and incident response framework. Today, several of the large insurance carriers I interviewed for this book are exploring underwriting AI risks as part of a subset of cyber insurance, or as a separate offering. Many are referring to it broadly as algorithmic risk in order to avoid distinctions between AI, machine learning, and other automation. "It's where the future is clearly going," said the chief innovation officer of one of these firms. "We quickly built a cyber insurance practice when the market needed it. Now, my fear is that we are just one disaster away from an AI risk practice. It would be better to start sooner, but the demand is not there."

Rounding off AI security as part of AI model risk management is model drift. Model drift is when a model's performance changes over time due to changes in the data it is receiving. Essentially, the model changes its behavior because it is learning new things. These changes are not always good and can create security gaps. Model drift requires active monitoring of an AI.

Model drift can occur in cases where the AI's input data changes over time relative to the data it was trained on without the model undergoing retraining on a periodic basis. For example, a stock trading AI might be trained in certain market conditions that prevailed in the years leading up to the AI's development. If the Federal Reserve takes an aggressive stance and intervenes in the market in new, or unforeseen ways, the new type of signals the model is getting might cause the model to become less effective. Models can also change their performance due to adversarial impacts, such as an attacker feeding a model bad information to change its performance.

Model drift can be monitored either directly or by proxy. In the case of direct monitoring, the AI engineer must collect and label new data to test the performance of the model on an ongoing basis. By doing this the AI engineer can observe if the quality of the model degrades over time and attempt to overcome this by performing new training cycles at an appropriate interval. This is typically the best way to monitor model drift. However, this approach comes with the overhead requirement of providing new labeled data on an ongoing basis. The easier, but less effective, solution is to monitor model drift by proxy of the input data. In this case the distribution of the input data is monitored over time; if changes are observed in the distribution of the input data it may mean that the model is becoming stale. The challenge is that without labels this cannot be tested explicitly, only inferred. A hybrid solution would monitor staleness using the proxy method and only apply the direct method in cases where the proxy indicates an increase in staleness. This has the benefit of avoiding unnecessary labeling efforts but the disadvantage of missing staleness in cases where the underlying distribution does not change—for example, if the distribution of the features and labels both change simultaneously in such a way as to mask the changes to the features alone.

Good security starts with active threat modeling and does not end, even while the model is in use. It is critical to perform ongoing security checks on models, red-team models even while in use, and maintain active monitoring of AI performance. Although security can never be 100 percent guaranteed, these measures will limit an organization's exposure to AI security risks.

ETHICS, LEGALITIES, RISK, AND COMPLIANCE

The wrong time to review both the underlying ethics and compliance requirements of a new AI system is at the end of its development time line. However, this is usually what happens at large enterprises. Instead of taking the time to assess these at the beginning, many times this is left to the compliance team only at the end. When this happens, massive time delays can take place as the compliance team needs to thoroughly research the use case, identify areas for legal, risk, and compliance review, and then thoroughly interrogate the AI and the data science team to ensure standards and ethical considerations are met. This time-consuming process is a big part of the flash-to-bang delays in deploying AI models.

Instead, prior to beginning a complete model development cycle, organizations should carefully consider these elements. Exceptions should be made for experimental models, of course, but even in experimental contexts careful ethical consideration must be made both to the underlying data and the use of the AI tool.

Legal and compliance tend to be the clearest elements of this section of AI model risk management to understand. Organizations in heavily regulated industries are already accustomed to rigorous compliance and legal reviews for new products and services. Smaller organizations or organizations in less regulated industries may need to rely on third-party legal or expert counsel when making their decisions. One increasingly critical risk is data privacy. California's recent consumer data protections and the European Union's General Data Protection Regulation (GDPR) have increased both the number of data compliance regulations as well as the costs associated with noncompliance. The ISO[23] and

NIST[24] trusted AI standards and frameworks are likely going to be the guiding principles for trusted AI in the near term, while industry-specific guidance will come from federal and state regulatory bodies. The U.S. Department of Transportation, for example, has started slow-rolling self-driving car guidance[25] and plans to release more as regulations are set.

Risk assessments for AI come in several forms. They are typically bucketed as performance and safety risk, legal risk, and reputational risk. In the context of AI, performance and safety risk includes the risk that something will go wrong and what the effect will be on the organization. AI applications in mission critical or public safety situations, such as weapons systems or self-driving vehicles, obviously have larger safety considerations than an AI that predicts warehouse supply. By determining performance risk early, organizations can also decide which safety and performance thresholds an AI has to meet prior to deployment. Legal risk looks both at what laws must be adhered to when both developing and using the AI. In assessing this risk, it is important to also evaluate worst-case scenarios for the AI's use, or misuse. Who carries legal risk can sometimes be murky. For example, if an AI model is purchased from a vendor and used in a system that fails, who is liable? Currently, it is case dependent and can rely on individual licensing agreements between the organizations. Finally, reputational risk is the risk to an organization's regulation if an AI goes awry or if knowledge of the AI's application is uncovered.

Unlike legal and compliance risks, ethical considerations are murkier. While many organizations tend to bucket ethical risk alongside reputation risk, I feel that this is the wrong approach. Firms and organizations that wish to turn certain critical thinking, and even decision-making, capabilities over to AI systems need to be aware not only about the multitude of performance and security risks such systems face. They also must take into account questions like: Who does this AI replace? Will any harm come from this AI's use? and Can my AI be abused? I am an AI optimist. I believe that the benefits from the technology can improve lives, decrease inequitable distributions, and lead to breakthroughs for society. But this is an optimist's view. Of course it is rosy. There is a very real world in which AI only fuels additional inequality, racial bias, joblessness, and inequitable distributions of wealth. And in fact, many believe this is a more likely outcome than my optimistic view. AI will not create either of these futures on its own. Humans must develop and deploy AI to meet their needs. Therefore, careful evaluation of AI ethics, including their codification into law when necessary, is the most powerful tool we have to prevent AI-enabled dystopian futures.

Ethical considerations are closely tied to underlying biases in the AI and data. The most important consideration when addressing bias is executive awareness and concern. Leaders must educate themselves about underlying data biases and the challenges it can bring to their successful implementation of AI. Organizations should also establish fact-based conversations and reporting around discussing data and AI bias. Many professionals feel it borders on the political and therefore avoid it. Such behavior can not only continue to harm vulnerable communities, but also leads to models of poor quality. Using modern model interpretability techniques can provide organizations with the facts they need about an AI's decisions to engage in transparent discussions. A successful AI model risk management process involves multiple touch points with many parts of an organization, and therefore should also help to diversify the professionals evaluating AI systems and their potential for bias. But this side effect of AI model risk management is not enough. Removing data and AI bias requires an active approach on behalf of data scientists and their organizations. As a technology community, we need to proactively work toward a more inclusive AI field.

AI ethics are closely tied to overall compliance. But it is important to remember that even if something is fully legal or compliant, an organization might decide to spike it due to ethical concerns. This is, and should be considered, a valid reason to not pursue AI projects. If done correctly, this decision can be made in a transparent and traceable manner that makes it clear to the organization why this decision was the right one.

Notes

Introduction
1. https://skylightcyber.com/2019/07/18/cylance-i-kill-you/
2. https://www.npr.org/2019/09/18/762046356/u-s-military-researchers-work-to-fix-easily-fooled-ai
3. https://arxiv.org/pdf/1712.03141.pdf
4. https://nicholas.carlini.com/writing/2019/all-adversarial-example-papers.html
5. https://www.bbc.com/news/technology-25506020

Chapter One
1. Daniel Crevier, *AI: The Tumultuous Search for Artificial Intelligence* (New York: Basic Books, 1993).
2. Marvin L. Minsky, *Computation: Finite and Infinite Machines* (Taipei: Central Book Co., 1967), p. 2, quoted in ibid., p. 109.
3. Crevier, *AI*, pp. 115–117.
4. Neil R. Smalheiser, "Walter Pitts," *Perspectives in Biology and Medicine* 43(2) (Winter 2000): 217–226.
5. David Poole, Alan Mackworth, and Randy Goebel, *Computational Intelligence: A Logical Approach* (New York: Oxford University Press, 1998).

Chapter Two
1. P. Surana, "Meta-Compilation of Language Abstractions." http://lispnyc.org/meeting-assets/2007-02-13_pinku/SuranaThesis.pdf. Archived from the original (PDF) on 2015-02-17. Retrieved 2008-03-17.
2. https://ai.stanford.edu/~zayd/why-is-machine-learning-hard.html

Chapter Three
1. Jeffrey Dastin, "Amazon scraps secret AI recruiting tool that showed bias against women," Reuters. October 9, 2018. https://www.reuters.com/article/us-amazon-com-jobs-automation-insight/amazon-scraps-secret-ai-recruiting-tool-that-showed-bias-against-women-idUSKCN1MK08G#:~:text=Amazon

%20scraps%20secret%20AI%20recruiting%20tool%20that%20showed%20
bias%20against%20women,-.

2. *British Medical Journal (Clinical Research Ed.)* 1988 Mar 5; 296(6623):
657–658. https://www.ncbi.nlm.nih.gov/pmc/articles/PMC2545288/?page=1

3. http://proceedings.mlr.press/v81/buolamwini18a/buolamwini18a.pdf

4. https://crashstats.nhtsa.dot.gov/Api/Public/ViewPublication/812115

5. https://www.transportation.gov/av/data

6. https://www.fastcompany.com/90269688/high-tech-redlining-ai-is-quietly
-upgrading-institutional-racism

7. https://www.propublica.org/article/machine-bias-risk-assessments-in
-criminal-sentencing

8. E. L. Barse, H. Kvarnström, and E. Jonsson, "Synthesizing test data for
fraud detection systems," Proceedings of the 19th Annual Computer Security
Applications Conference. IEEE, 2003.

9. https://www.aaai.org/ojs/index.php/AAAI/article/view/4777

10. https://fcw.com/articles/2020/02/24/dod-ai-policy-memo-williams.aspx

CHAPTER FOUR

1. https://arxiv.org/pdf/1707.08945.pdf

CHAPTER FIVE

1. https://arxiv.org/abs/1412.6572

2. https://arxiv.org/abs/1312.6199

3. https://arxiv.org/abs/1412.6572

4. https://arxiv.org/abs/1412.6572

5. https://arxiv.org/abs/1608.07690

6. https://arxiv.org/abs/1804.11285

7. https://arxiv.org/abs/1805.10204

8. https://arxiv.org/abs/1905.02175

9. If a reader would like to bend their mind a little more to understand this
concept, this blog does a good job at summing it up: http://gradientscience
.org/adv/

10. https://arxiv.org/abs/1705.07263

11. Paul Bischoff, "Surveillance camera statistics: which cities have the most
CCTV cameras?" https://www.comparitech.com/vpn-privacy/the-worlds
-most-surveilled-cities/ Archived from the original on 2019-10-09.
Retrieved 2019-11-14.

12. The world's most-surveilled cities. *Comparitech.* Archived from the original
on 2019-10-09. Retrieved 2019-11-14.

13. M. Sharif, S. Bhagavatula, L. Bauer, and M. K. Reiter, "Accessorize to a
crime: Real and stealthy attacks on state-of-the-art face recognition." In

Proceedings of the 2016 ACM SIGSAC Conference on Computer and Communications Security, October 24–28, 2016, Vienna, Austria; pp. 1528–40.

14. Some researchers disagree with this definition, calling these attacks Black-Box attacks. I believe the distinction is warranted as some level of information is provided.

15. https://arxiv.org/abs/1708.03999
16. https://arxiv.org/abs/1802.05666
17. https://arxiv.org/abs/1804.08598
18. https://arxiv.org/abs/1712.04248
19. https://arxiv.org/abs/1712.02779
20. https://arxiv.org/abs/1807.01697
21. https://arxiv.org/abs/1901.10513
22. https://www.businessinsider.com/the-secret-bin-laden-training-facility-2012-10
23. https://www.theatlantic.com/international/archive/2012/10/satellite-images-capture-cias-secret-bin-laden-training-facility/322676/
24. https://www.usenix.org/system/files/sec19-demontis.pdf
25. https://arxiv.org/abs/1609.02943
26. https://arxiv.org/abs/2004.06954
27. https://arxiv.org/pdf/1910.07067.pdf

Chapter Six

1. https://nds2.ccs.neu.edu/papers/aml_aisec2014.pdf
2. https://arxiv.org/abs/1804.07933
3. http://www0.cs.ucl.ac.uk/staff/B.Karp/polygraph-oakland2005.pdf
4. http://people.ischool.berkeley.edu/~tygar/papers/SML/IMC.2009.pdf
5. https://link.springer.com/chapter/10.1007/11856214_5
6. https://link.springer.com/chapter/10.1007/11856214_4
7. https://link.springer.com/chapter/10.1007/978-3-540-74320-0_13
8. https://towardsdatascience.com/poisoning-attacks-on-machine-learning-1ff247c254db
9. https://ieeexplore.ieee.org/stamp/stamp.jsp?arnumber=8685687
10. https://arxiv.org/abs/1708.08689
11. https://arxiv.org/abs/1706.03691
12. https://arxiv.org/abs/1703.01340
13. James Vincent, "Twitter taught Microsoft's AI chatbot to be a racist asshole in less than a day, *The Verge,* March 24, 2016. https://www.theverge.com/2016/3/24/11297050/tay-microsoft-chatbot-racist
14. https://arxiv.org/pdf/1710.08864.pdf
15. https://pdfs.semanticscholar.org/5f25/7ca18a92c3595db3bda3224927ec494003a5.pdf
16. https://arxiv.org/abs/1902.06531

CHAPTER SEVEN
1. https://arxiv.org/abs/1610.05820
2. https://arxiv.org/abs/1610.05820
3. https://arxiv.org/pdf/1811.00513.pdf
4. https://arxiv.org/abs/1708.06145
5. https://www.usenix.org/system/files/conference/usenixsecurity14/sec14 -paper-fredrikson-privacy.pdf
6. https://www.cs.cmu.edu/~mfredrik/papers/fjr2015ccs.pdf
7. https://arxiv.org/pdf/1802.08232.pdf
8. https://arxiv.org/abs/1609.02943
9. https://arxiv.org/abs/2005.10284

CHAPTER EIGHT
1. https://adversarial-attacks.net/
2. https://www.fastcompany.com/90240975/alexa-can-be-hacked-by-chirping -birds
3. Scott M. Lundberg and Su-In Lee, "A Unified Approach to Interpreting Model Predictions," *Advances in Neural Information Processing Systems 30*. https://proceedings.neurips.cc/paper/2017/file/8a20a8621978632d 76c43dfd28b67767-Paper.pdf. Retrieved 2020-03-13.

CHAPTER NINE
1. https://towardsdatascience.com/human-interpretable-machine-learning -part-1-the-need-and-importance-of-model-interpretation-2ed758f5f476
2. https://www.kdd.org/kdd2016/papers/files/rfp0573-ribeiroA.pdf
3. https://www.nga.mil/About/History/NGAinHistory/Pages/NCEOpens .aspx
4 .https://www.nga.mil/news/NGA_Tech_Focus_Areas_chart_path_for _geospatial_tec.html

CHAPTER TEN
1. https://www.reuters.com/article/us-cyber-deepfake-activist/deepfake-used -to-attack-activist-couple-shows-new-disinformation-frontier-idUSKCN24G 15E
2. https://www.technologyreview.com/2018/02/21/145289/the-ganfather-the -man-whos-given-machines-the-gift-of-imagination/

CHAPTER ELEVEN
1. https://www.microsoft.com/en-us/securityengineering/sdl/
2. https://www.microsoft.com/en-us/securityengineering/sdl/

3. Mark Savinson, "Artificial Intelligence in Sales: Is It Worth the Investment?" *Forbes*, April 17, 2019. https://www.forbes.com/sites/forbescoachescouncil /2019/04/17/artificial-intelligence-in-sales-is-it-worth-the-investment/?sh =24f82df4201e.

4. https://www.darkreading.com/vulnerabilities---threats/average-cost-of-a -data-breach-$116m/a/d-id/1338121

5. https://www.scasecurity.com/cost-of-a-data-breach/#:~:text=According%20 to%20the%202019%20Cost,an%20average%20of%20%243.92%20million.

6. https://www.penguinrandomhouse.com/books/534564/red-teaming-by -bryce-g-hoffman/

7. https://adversarial-robustness-toolbox.readthedocs.io/en/latest/#:~:text =Adversarial%20Robustness%20Toolbox%20(ART)%20is,Poisoning%2C%20 Extraction%2C%20and%20Inference.

CHAPTER TWELVE

1. https://www.federalreserve.gov/supervisionreg/srletters/sr1107.htm

2. https://www.federalreserve.gov/supervisionreg/srletters/sr1107.htm

3. https://www.federalreserve.gov/boarddocs/supmanual/trading/trading.pdf

4. https://www.fdic.gov/news/financial-institution-letters/2017/fil17022.html

5. https://www.fdic.gov/news/financial-institution-letters/2017/fil17022.html

6. https://www.occ.gov/news-issuances/bulletins/2011/bulletin-2011-12.html

7. https://www.occ.gov/news-issuances/bulletins/2011/bulletin-2011-12.html

8. https://www.disruptordaily.com/ai-challenges-insurance/

9. https://breakingdefense.com/2018/06/joint-artificial-intelligence-center -created-under-dod-cio/

10. https://www.defense.gov/Explore/News/Article/Article/1254719/project -maven-to-deploy-computer-algorithms-to-war-zone-by-years-end/

11. https://www.rand.org/news/press/2019/12/17.html

12. It should be noted that many large, digital native companies such as Uber, Amazon, and Alphabet seem to have robust internally created AI model validation and model risk management solutions already.

13. https://www.gartner.com/en/documents/3899783/anticipate-data-manipu lation-security-risks-to-ai-pipeli

14. https://hbr.org/2020/04/the-case-for-ai-insurance

15. https://arxiv.org/abs/2002.08347

16. https://arxiv.org/pdf/1810.10731.pdf

17. https://medium.com/@karpathy/software-2-0-a64152b37c35

18. https://www.zdnet.com/article/googles-best-image-recognition-system -flummoxed-by-fakes/

19. https://spectrum.ieee.org/cars-that-think/transportation/self-driving/three -small-stickers-on-road-can-steer-tesla-autopilot-into-oncoming-lane

20. https://www.washingtonpost.com/local/trafficandcommuting/pedestrian -in-self-driving-uber-collision-probably-would-have-lived-if-braking-feature

-hadnt-been-shut-off-ntsb-finds/2019/11/05/7ec83b9c-ffeb-11e9-9518-1e76
abc088b6_story.html

21. https://www.iso.org/committee/6794475.html
22. https://www.nist.gov/topics/artificial-intelligence/ai-standards
23. https://www.iso.org/committee/6794475.html
24. https://www.nist.gov/topics/artificial-intelligence/ai-standards
25. https://www.transportation.gov/av/3

Bibliography

Ackerman, Evan. "Three Small Stickers in Intersection Can Cause Tesla
Autopilot to Swerve into Wrong Lane." IEEE Spectrum, June 24, 2021.
https://spectrum.ieee.org/cars-that-think/transportation/self-driving
/three-small-stickers-on-road-can-steer-tesla-autopilot-into-oncoming
-lane.

Adversarial Robustness Toolbox. "Welcome to the Adversarial Robustness Toolbox." Adversarial Robustness Toolbox 1.7.2 documentation.
Accessed September 9, 2021. https://adversarial-robustness-toolbox.
readthedocs.io/en/latest/#:~:text=Adversarial%20Robustness%20Tool
box%20(ART)%20is,Poisoning%2C%20Extraction%2C%20and%20
Inference.

Angwin, Julia, and Jeff Larson. "Machine Bias." ProPublica, May 23, 2016.
https://www.propublica.org/article/machine-bias-risk-assessments
-in-criminal-sentencing.

Barse, E. L., H. Kvarnstrom, and E. Jonsson, "Synthesizing test data for fraud
detection systems." 19th Annual Computer Security Applications Conference, 2003. Proceedings., 2003, pp. 384–394.

Biggio, Battista, and Fabio Roli. "Wild Patterns: Ten Years After the Rise of
Adversarial Machine Learning." Arxiv, July 19, 2018. https://arxiv.org
/pdf/1712.03141.pdf.

Bischoff, Paul. "Surveillance Camera Statistics: Which City Has the Most
CCTV Cameras?" Comparitech, June 8, 2021. https://www.comparitech
.com/vpn-privacy/the-worlds-most-surveilled-cities/.

Brendel, Wieland, Jonas Rauber, and Matthias Bethge. "Decision-based
adversarial attacks: Reliable attacks against black-box machine learning
models." arXiv preprint arXiv:1712.04248 (2017).

Bubeck, Sébastien, Yin Tat Lee, Eric Price, and Ilya Razenshteyn. "Adversarial
examples from computational constraints." In International Conference
on Machine Learning, pp. 831–840. PMLR, 2019.

Buolamwini, Joy, and Timnit Gebru. "Gender Shades: Intersectional Accuracy
Disparities in Commercial Gender Classification." *Proceedings of Machine*

Learning Research 81 (February 23, 2018): 77–91. https://doi.org/http:// proceedings.mlr.press/v81/buolamwini18a/buolamwini18a.pdf.

Carlini, Nicholas. "A Complete List of All (ArXiv) Adversarial Example Papers." June 15, 2019. https://nicholas.carlini.com/writing/2019/all -adversarial-example-papers.html.

Carlini, Nicholas, and David Wagner. "Adversarial examples are not easily detected: Bypassing ten detection methods." In Proceedings of the 10th ACM workshop on artificial intelligence and security, pp. 3–14. 2017.

Carlini, Nicholas, Chang Liu, Úlfar Erlingsson, Jernej Kos, and Dawn Song. "The secret sharer: Evaluating and testing unintended memorization in neural networks." In 28th {USENIX} Security Symposium ({USENIX} Security 19), pp. 267–284. 2019.

Chen, Pin-Yu, Huan Zhang, Yash Sharma, Jinfeng Yi, and Cho-Jui Hsieh. "Zoo: Zeroth order optimization based black-box attacks to deep neural networks without training substitute models." In Proceedings of the 10th ACM workshop on artificial intelligence and security, pp. 15–26. 2017.

Chiappa, Silvia. 2019. "Path-Specific Counterfactual Fairness." Proceedings of the AAAI Conference on Artificial Intelligence 33 (01):7801–8. https:// doi.org/10.1609/aaai.v33i01.33017801.

Chung, Simon P., and Aloysius K. Mok. "Advanced allergy attacks: Does a corpus really help?." In International Workshop on Recent Advances in Intrusion Detection, pp. 236–255. Springer, Berlin, Heidelberg, 2007.

Chung, Simon P., and Aloysius K. Mok. "Allergy attack against automatic signature generation." In International Workshop on Recent Advances in Intrusion Detection, pp. 61–80. Springer, Berlin, Heidelberg, 2006.

Crevier, Daniel. *AI: The Tumultuous History of the Search for Artificial Intelligence.* New York: Basic Books, 1993.

Dastin, Jeffrey. "Amazon Scraps Secret AI Recruiting Tool That Showed Bias against Women." Thomson Reuters, October 10, 2018. https://www .reuters.com/article/us-amazon-com-jobs-automation-insight/amazon -scraps-secret-ai-recruiting-tool-that-showed-bias-against-women-id USKCN1MK08G.

Demontis, Ambra, Marco Melis, Maura Pintor, Matthew Jagielski, Battista Biggio, Alina Oprea, Cristina Nita-Rotaru, and Fabio Roli. "Why do adversarial attacks transfer? Explaining transferability of evasion and poisoning attacks." In 28th {USENIX} Security Symposium ({USENIX} Security 19), pp. 321–338. 2019.

Diaz, Jesus. "Alexa Can Be Hacked—by Chirping Birds." Fast Company, September 28, 2018. https://www.fastcompany.com/90240975/alexa -can-be-hacked-by-chirping-birds.

Enam, S. Zayd. "Why Is Machine Learning 'Hard'?" Zayd's Blog, November 10, 2016. https://ai.stanford.edu/~zayd/why-is-machine-learning-hard .html

Engstrom, Logan, Brandon Tran, Dimitris Tsipras, Ludwig Schmidt, and Aleksander Madry. "Exploring the landscape of spatial robustness." In *International Conference on Machine Learning*, pp. 1802–1811. PMLR, 2019.

Erwin, Sandra. "NGA Official: Artificial Intelligence Is Changing Everything, "We Need a Different Mentality." *SpaceNews,* May 13, 2018. https://spacenews.com/nga-official-artificial-intelligence-is-changing-every thing-we-need-a-different-mentality/.

Eykholt, Kevin, I. Evtimov, Earlence Fernandes, Bo Li, Amir Rahmati, Chaowei Xiao, Atul Prakash, T. Kohno, and D. Song. "Robust Physical-World Attacks on Deep Learning Visual Classification." 2018 IEEE/CVF Conference on Computer Vision and Pattern Recognition (2018): 1625–1634.

Fast Company. "High-Tech Redlining: AI Is Quietly Upgrading Institutional Racism." Fast Company, November 20, 2018. https://www.fastcompany.com/90269688/high-tech-redlining-ai-is-quietly-upgrading-institution al-racism.

Federal Reserve. "Board of Governors of the Federal Reserve System." Supervisory Letter SR 11–7 on guidance on Model Risk Management, April 4, 2011. Accessed September 9, 2021. https://www.federalreserve.gov /supervisionreg/srletters/sr1107.htm.

Federal Reserve Board. "Trading and Capital-Markets Activities Manual." February 1998. https://www.federalreserve.gov/boarddocs/supmanual /trading/trading.pdf.

Ford, Nic, Justin Gilmer, Nicolas Carlini, and Dogus Cubuk. "Adversarial examples are a natural consequence of test error in noise." arXiv preprint arXiv:1901.10513 (2019).

Fredrikson, Matt, Somesh Jha, and Thomas Ristenpart. "Model inversion attacks that exploit confidence information and basic countermeasures." In Proceedings of the 22nd ACM SIGSAC conference on computer and communications security, pp. 1322–1333. 2015.

Fredrikson, Matthew, Eric Lantz, Somesh Jha, Simon Lin, David Page, and Thomas Ristenpart. "Privacy in pharmacogenetics: An end-to-end case study of personalized warfarin dosing." In 23rd {USENIX} Security Symposium ({USENIX} Security 14), pp. 17–32. 2014.

Freedberg, Sydney J. "Joint Artificial Intelligence Center Created under DOD CIO." Breaking Defense, July 22, 2021. https://breakingdefense .com/2018/06/joint-artificial-intelligence-center-created-under-dod-cio/.

Gao, Yansong, Change Xu, Derui Wang, Shiping Chen, Damith C. Ranas-inghe, and Surya Nepal. "Strip: A defence against Trojan attacks on deep neural networks." In Proceedings of the 35th Annual Computer Security Applications Conference, pp. 113–125. 2019.

Gartner, Inc. "Anticipate Data Manipulation Security Risks to AI Pipelines." Gartner. Accessed September 9, 2021. https://www.gartner.com/en/docu ments/3899783/anticipate-data-manipulation-security-risks-to-ai-pipeli.

Giles, Martin. "The Ganfather: The Man Who's given Machines the Gift of Imagination." *MIT Technology Review*, April 2, 2020. https://www.tech nologyreview.com/2018/02/21/145289/the-ganfather-the-man-whos -given-machines-the-gift-of-imagination/.

Goodfellow, Ian J., Jonathon Shlens, and Christian Szegedy. "Explaining and harnessing adversarial examples." arXiv preprint arXiv:1412.6572 (2014).

Gu, Tianyu, Kang Liu, Brendan Dolan-Gavitt, and Siddharth Garg. "Badnets: Evaluating backdooring attacks on deep neural networks." *IEEE Access* 7 (2019): 47230–47244.

Hendrycks, Dan, and Thomas G. Dietterich. "Benchmarking neural network robustness to common corruptions and surface variations." arXiv preprint arXiv:1807.01697 (2018).

Hoffman, Bryce G. *Red Teaming: How Your Business Can Conquer the Competition by Challenging Everything*. New York: Crown Business, 2017.

Hudson, John. "Satellite Images of the CIA'S Secret Bin LADEN Training Facility." *The Atlantic*, October 30, 2013. https://www.theatlantic.com /international/archive/2012/10/satellite-images-capture-cias-secret -bin-laden-training-facility/322676/.

Ilyas, Andrew, Logan Engstrom, Anish Athalye, and Jessy Lin. "Black-box adversarial attacks with limited queries and information." In International Conference on Machine Learning, pp. 2137–2146. PMLR, 2018.

Ilyas, Andrew, Shibani Santurkar, Dimitris Tsipras, Logan Engstrom, Brandon Tran, and Aleksander Madry. "Adversarial examples are not bugs, they are features." arXiv preprint arXiv:1905.02175 (2019).

ISO. "ISO/IEC JTC 1/SC 42—Artificial Intelligence." ISO, September 9, 2021. https://www.iso.org/committee/6794475.html.

Karpathy, Andrej. "Software 2.0." *Medium*, March 13, 2021. https://medium .com/@karpathy/software-2-0-a64152b37c35.

Kelion, Leo. "Cryptolocker Ransomware Has 'Infected about 250,000 PCS.'" BBC News. BBC, December 24, 2013. https://www.bbc.com/news/tech nology-25506020.

Kelley, Michael B. "Bing Maps Show The CIA'S Secret Bin LADEN Training Facility in North Carolina." Business Insider, October 9, 2012. https:// www.businessinsider.com/the-secret-bin-laden-training-facility-2012-10.

Kumar, Ram Shankar Siva, David R. O'Brien, Kendra Albert, and Salome Vilojen. "Law and Adversarial Machine Learning." arXiv preprint arXiv:1810.10731 (2018).

Laris, Michael. "Pedestrian in Self-Driving Uber Crash Probably Would Have Lived If Braking Feature Hadn't Been Shut off, NTSB Documents Show." *Washington Post*, November 6, 2019. https://www.washingtonpost

.com/local/trafficandcommuting/pedestrian-in-self-driving-uber-collision
-probably-would-have-lived-if-braking-feature-hadnt-been-shut-off
-ntsb-finds/2019/11/05/7ec83b9c-ffeb-11e9-9518-1e76abc088b6
_story.html.

Lauren C. Williams. "DOD Releases First AI Ethics Principles, but There's Work Left to on Implementation." FCW. February 24, 2020. Accessed September 9, 2021. https://fcw.com/articles/2020/02/24/dod-ai-policy -memo-williams.aspx.

Lei, Yusi, Sen Chen, Lingling Fan, Fu Song, and Yang Liu. "Advanced evasion attacks and mitigations on practical ML-based phishing website classifiers." arXiv preprint arXiv:2004.06954 (2020).

Liu, Yingqi, Shiqing Ma, Yousra Aafer, Wen-Chuan Lee, Juan Zhai, Weihang Wang, and Xiangyu Zhang. "Trojaning attack on neural networks." Department of Computer Science Technical Reports. Department of Computer Science. Purdue University. (2017).

Lowry, Stella, and Gordon Macpherson. "A Blot on the Profession." *British Medical Journal* 296, no. 6623 (March 5, 1988): 657–58. https://doi.org /https://www.ncbi.nlm.nih.gov/pmc/articles/PMC2545288/?page=1.

Lundberg, Scott M., and Su-In Lee. "A unified approach to interpreting model predictions." In Proceedings of the 31st international conference on neural information processing systems, pp. 4768–4777. 2017.

McCulloch, Warren, and Walter Pitts . "A Logical Calculus of the Ideas Immanent in Nervous Activity." *Bulletin of Mathematical Physics* 5 (1943): 115–33. https://doi.org/https://homeweb.csulb.edu/~cwallis/382/readings /482/mccolloch.logical.calculus.ideas.1943.pdf.

Microsoft. "Microsoft Security Development Lifecycle." Accessed September 9, 2021. https://www.microsoft.com/en-us/securityengineering/sdl/.

Miller, David J., Zhen Xiang, and George Kesidis. "Adversarial learning in statistical classification: A comprehensive review of defenses against attacks." arXiv preprint arXiv:1904.06292 (2019).

Minsky, Marvin L. *Computation: Finite and Infinite Machines.* Taipei: Central Book Co., 1967.

Mire, Sam. "What Are the Challenges to AI Adoption in Insurance? 11 Experts Share Their Insights." *Disruptor Daily*, September 25, 2019. https://www.disruptordaily.com/ai-challenges-insurance/.

Moisejevs, Ilja. "Poisoning Attacks on Machine Learning." Towards Data Science, July 15, 2019. https://towardsdatascience.com/poisoning-attacks -on-machine-learning-1ff247c254db.

Muñoz-González, Luis, Battista Biggio, Ambra Demontis, Andrea Paudice, Vasin Wongrassamee, Emil C. Lupu, and Fabio Roli. "Towards poisoning of deep learning algorithms with back-gradient optimization." In Proceedings of the 10th ACM Workshop on Artificial Intelligence and Security, pp. 27–38. 2017.

National Highway Traffic Safety Administration. "Critical Reasons for Crashes Investigated in the National Motor Vehicle Crash Causation Survey." Traffic Safety Facts. U.S. Department of Transportation, February 2015. Critical Reasons for Crashes Investigated in the National Motor Vehicle Crash Causation Survey.

Newell, Andrew, Rahul Potharaju, Luojie Xiang, and Cristina Nita-Rotaru. "On the practicality of integrity attacks on document-level sentiment analysis." In Proceedings of the 2014 Workshop on Artificial Intelligent and Security Workshop, pp. 83–93. 2014.

Newsome, James, Brad Karp, and Dawn Song. "Paragraph: Thwarting signature learning by training maliciously." In International Workshop on Recent Advances in Intrusion Detection, pp. 81–105. Springer, Berlin, Heidelberg, 2006.

Newsome, James, Brad Karp, and Dawn Song. "Polygraph: Automatically generating signatures for polymorphic worms." In 2005 IEEE Symposium on Security and Privacy (S&P'05), pp. 226–241. IEEE, 2005.

NGA. "Associated Document(s)." NGA Tech Focus Areas chart path for geospatial technologies | National Geospatial-Intelligence Agency, April 29, 2020. https://www.nga.mil/news/NGA_Tech_Focus_Areas_chart_path _for_geospatial_tec.html.

FDIC. "Adoption of Supervisory Guidance on Model Risk Management." Financial Institution Letter. FIL-22-2017. Washington, D.C., June 7, 2017.

NIST. "AI Standards: Federal Engagement." NIST, August 10, 2021. https:// www.nist.gov/topics/artificial-intelligence/ai-standards.

OCC. "Sound Practices for Model Risk Management: Supervisory Guidance on Model Risk Management." OCC, April 4, 2011. https://www.occ.gov /news-issuances/bulletins/2011/bulletin-2011-12.html.

Pautov, Mikhail, Grigorii Melnikov, Edgar Kaziakhmedov, Klim Kireev, and Aleksandr Petiushko. "On adversarial patches: Real-world attack on arcface-100 face recognition system." In 2019 International Multi-Conference on Engineering, Computer and Information Sciences (SIBIRCON), pp. 0391–0396. IEEE, 2019.

Pellerin, Cheryl. "Project Maven to Deploy Computer Algorithms to War Zone by Year's End." U.S. Department of Defense. Accessed September 9, 2021. https://www.defense.gov/Explore/News/Article/Article/1254719 /project-maven-to-deploy-computer-algorithms-to-war-zone-by-years -end/.

Poole, David L., Alan Mackworth, and Randy Goebel. *Computational Intelligence: A Logical Approach*. New York: Oxford University Press, 1998.

Pyrgelis, Apostolos, Carmela Troncoso, and Emiliano De Cristofaro. "Knock knock, who's there? Membership inference on aggregate location data." arXiv preprint arXiv:1708.06145 (2017).

Rahnama, Arash, and Andrew Tseng. "An Adversarial Approach for Explaining the Predictions of Deep Neural Networks." In Proceedings of the IEEE/ CVF Conference on Computer Vision and Pattern Recognition, pp. 3253–3262. 2021.

Ray, Tiernan. "Google's Image Recognition AI Fooled by New Tricks." ZDNet, November 30, 2018. https://www.zdnet.com/article/googles-best-image -recognition-system-flummoxed-by-fakes/.

Ren, Kui, Tianhang Zheng, Zhan Qin, and Xue Liu. "Adversarial attacks and defenses in deep learning." *Engineering* 6, no. 3 (2020): 346–360.

Ribeiro, Marco Tulio, Sameer Singh, and Carlos Guestrin. "'Why should i trust you?' Explaining the predictions of any classifier." In Proceedings of the 22nd ACM SIGKDD international conference on knowledge discovery and data mining, pp. 1135–1144. 2016.

Rubinstein, Benjamin I.P., Blaine Nelson, Ling Huang, Anthony D. Joseph, Shing-hon Lau, Satish Rao, Nina Taft, and J. Doug Tygar. "Antidote: understanding and defending against poisoning of anomaly detectors." In Proceedings of the 9th ACM SIGCOMM Conference on Internet Measurement, pp. 1–14. 2009.

Sarkar, Dipanjan (DJ). "The Importance of Human Interpretable Machine Learning." Towards Data Science, December 13, 2018. https://towards datascience.com/human-interpretable-machine-learning-part-1-the -need-and-importance-of-model-interpretation-2ed758f5f476?gi=2d2e 69acf123.

Satter, Raphael. "Deepfake Used to Attack ACTIVIST Couple Shows New Disinformation Frontier." Reuters, July 15, 2020. https://www.reuters .com/article/us-cyber-deepfake-activist/deepfake-used-to-attack-activist -couple-shows-new-disinformation-frontier-idUSKCN24G15E.

Savinson, Mark. "Artificial Intelligence in Sales: Is It Worth the Investment?" *Forbes*, April 17, 2019. https://www.forbes.com/sites/forbescoaches council/2019/04/17/artificial-intelligence-in-sales-is-it-worth-the -investment/?sh=24f82df4201e.

Schmidt, Ludwig, Shibani Santurkar, Dimitris Tsipras, Kunal Talwar, and Aleksander Mądry. "Adversarially robust generalization requires more data." arXiv preprint arXiv:1804.11285 (2018).

Schönherr, Lea, Katharina Kohls, Steffen Zeiler, Thorsten Holz, and Dorothea Kolossa. "Adversarial Attacks Against ASR Systems VIA Psychoacoustic HIDING." Adversarial Attacks. Accessed September 9, 2021. https:// adversarial-attacks.net/.

Shankar, Ram, Siva Kumar, and Frank Nagle. "The Case for AI Insurance." *Harvard Business Review*, May 5, 2020. https://hbr.org/2020/04/the-case -for-ai-insurance.

Sharif, Mahmood, Sruti Bhagavatula, Lujo Bauer, and Michael K. Reiter. "Accessorize to a crime: Real and stealthy attacks on state-of-the-art face recognition." In Proceedings of the 2016 acm sigsac conference on computer and communications security, pp. 1528–1540. 2016.

Shokri, Reza, Marco Stronati, and Vitaly Shmatikov. "Membership Inference Attacks against Machine Learning Models. CoRR abs/1610.05820 (2016)." arXiv preprint arXiv:1610.05820 (2016).

Skylight Cyber. "Cylance, I Kill You!" July 18, 2019. https://skylightcyber.com /2019/07/18/cylance-i-kill-you/.

Song, Congzheng, and Vitaly Shmatikov. "Auditing data provenance in text-generation models." In Proceedings of the 25th ACM SIGKDD International Conference on Knowledge Discovery & Data Mining, pp. 196–206. 2019.

Steinhardt, Jacob, Pang Wei Koh, and Percy Liang. "Certified defenses for data poisoning attacks." In Proceedings of the 31st International Conference on Neural Information Processing Systems, pp. 3520–3532. 2017.

Su, Jiawei, Danilo Vasconcellos Vargas, and Kouichi Sakurai. "One pixel attack for fooling deep neural networks." *IEEE Transactions on Evolutionary Computation* 23, no. 5 (2019): 828–841.

Surana, Pankaj. "Meta-Compilation of Language Abstractions." http://lispnyc .org/meeting-assets/2007-02-13_pinku/SuranaThesis.pdf.

Szegedy, Christian, Wojciech Zaremba, Ilya Sutskever, Joan Bruna, Dumitru Erhan, Ian Goodfellow, and Rob Fergus. "Intriguing properties of neural networks." arXiv preprint arXiv:1312.6199 (2013).

Tanay, Thomas, and Lewis Griffin. "A boundary tilting persepective on the phenomenon of adversarial examples." arXiv preprint arXiv:1608.07690 (2016).

Tarraf, Danielle C. "Pentagon's Ambitious Vision and Strategy for AI Not Yet Backed by Sufficient Visibility or Resources." RAND Corporation, December 17, 2019. https://www.rand.org/news/press/2019/12/17.html.

Temple-Raston, Dina. "Computer Scientists Work to Fix Easily Fooled Ai." ' NPR, September 18, 2019. https://www.npr.org/2019/09/18/762046356 /u-s-military-researchers-work-to-fix-easily-fooled-ai.

Tramèr, Florian, Fan Zhang, Ari Juels, Michael K. Reiter, and Thomas Ristenpart. "Stealing machine learning models via prediction apis." In 25th {USENIX} Security Symposium ({USENIX} Security 16), pp. 601–618. 2016.

Tramer, Florian, Nicholas Carlini, Wieland Brendel, and Aleksander Madry. "On adaptive attacks to adversarial example defenses." arXiv preprint arXiv:2002.08347 (2020).

Uesato, Jonathan, Brendan O'donoghue, Pushmeet Kohli, and Aaron Oord. "Adversarial risk and the dangers of evaluating against weak attacks." In International Conference on Machine Learning, pp. 5025–5034. PMLR, 2018.

U.S. Department of Transportation. "Data for Automated Vehicle Integration (DAVI)." May 20, 2020. https://www.transportation.gov/av/data.

U.S. Department of Transportation. "Preparing for the Future of Transportation: Automated Vehicles 3.0." https://www.transportation.gov/av/3.

Vincent, James. "Twitter Taught Microsoft's AI Chatbot to Be a Racist Asshole in Less than a Day." *The Verge*, March 24, 2016. https://www.theverge.com/2016/3/24/11297050/tay-microsoft-chatbot-racist.

Wilczek, Marc. "Average Cost of a Data Breach: $116m." Dark Reading, June 29, 2021. https://www.darkreading.com/vulnerabilities---threats/average-cost-of-a-data-breach-$116m/a/d-id/1338121.

Xiao, Huang, Battista Biggio, Gavin Brown, Giorgio Fumera, Claudia Eckert, and Fabio Roli. "Is feature selection secure against training data poisoning?" In International Conference on Machine Learning, pp. 1689–1698. PMLR, 2015.

Yang, Chaofei, Qing Wu, Hai Li, and Yiran Chen. "Generative poisoning attack method against neural networks." arXiv preprint arXiv:1703.01340 (2017).

INDEX

Page numbers in italics refer to figures.

About the Author

Davey Gibian is a technologist and artificial intelligence practitioner. His career has spanned Wall Street, the White House, and active war zones as he has brought cutting-edge data science tools to solve hard problems. He has built two start-ups, Calypso AI and OMG, was a White House Presidential Innovation Fellow for Artificial Intelligence and Cybersecurity, and helped scale Palantir Technologies. He holds patents in machine learning and an undergraduate degree from Columbia University. Davey served in the U.S. Air Force and currently resides in New York City.

CPSIA information can be obtained
at www.ICGtesting.com
Printed in the USA
BVHW042303040422
633381BV00002B/2